THE SEVEN MAGNIFICENT

AF655219

THE SEVEN MAGNIFICENT

Simple, Quick and Cost Effective Tools for Continuous Improvement

Prof. A. N. Nankana

Patron, Quality Improvement Mission

NEW AGE

NEW AGE INTERNATIONAL (P) LIMITED, PUBLISHERS

New Delhi • Bangalore • Chennai • Cochin • Guwahati • Hyderabad
Jalandhar • Kolkata • Lucknow • Mumbai • Ranchi

Copyright © 2005, New Age International (P) Ltd., Publishers

NEW AGE INTERNATIONAL (P) LIMITED, PUBLISHERS
4835/24, Ansari Road, Daryaganj,
New Delhi - 110 002
Visit us at : www.newagepublishers.com

Offices at :
Bangalore, Chennai, Cochin, Guwahati, Hyderabad, Jalandhar,
Kolkata, Lucknow, Mumbai and Ranchi

This book or any part thereof may not be reproduced in any form without the written permission of the publisher.

This book cannot be sold outside the country to which it is consigned by the publisher without the prior permission of the publisher.

Rs. 95.00

ISBN : 81-224-1600-4

1 2 3 4 5 6 7 8 9 10

Published by New Age International (P) Ltd.,
4835/24, Ansari Road, Daryaganj, New Delhi-110 002 and
printed in India at Taj Press, Delhi
typesetter Laser Craft, Delhi

Dedicated to the cause of prosperity
released during 2004

the 111th birth anniversary year
of Prof. P.C. Mahalonobis
the Founder Director of
Indian Statistical Institute
the temple of my learning

the centenary year of Dr. J.M. Juran
the world quality Guru

the 84th year Dr. C.R. Rao
the best living statistician
and my esteemed guru.

FOREWORD

Indian industry has made tremendous progress in the last three to four decades. It has established manufacturing capacity in wide ranging products and technologies including those in high-tech areas like space, computers and biotech. In terms of quantitative production India ranks among the top few countries. However, on the quality aspect the picture is entirely different. In terms of exports which is true indicator of quality and cost competitiveness, India ranks even lower than small countries in South East Asia which started industrialization much later than India. This is in spite of the solid advantages of vast natural resources, highly competent managerial and technical manpower and low cost of labor.

Main reasons for this seeming paradox is that managements of Indian industries pay only lip service to quality management under mistaken belief that emphasis on quality would add to costs. This myth has been shattered by a number of controlled studies in Europe and Japan. The only way Indian industry can take its rightful place is through systematically planned quality improvement projects which can significantly improve quality conformance and also lead to cost reduction. It is here that the well-tried seven statistical tools can be used to make tangible improvements in quality and cost saving through elimination of wastes of all types.

One reason why these statistical tools are not fully utilized even in professionally managed companies is that these are perceived to be difficult to understand and apply because of advanced mathematics and statistics. Prof Nankana has done commendable service to industry by explaining these tools in simple language which even persons with high school background can follow. He has not stopped at explaining the concepts of these tools, but has gone much further by providing the methodology of their deployment. To convince the skeptics in industry, he has backed up application methodology with real life case studies based on his vast experience of consultancy with industries in different sectors.

I have had long association with Prof Nankana going back to our college days. During my tenure as Director General of Defence Quality Assurance and Bureau of Indian Standards, I had the opportunity to work closely with Prof Nankana on policy issues for quality promotion and development of National Standards on Quality Management. He always had down to earth shop floor approach and put across his views with passion, which cannot fail to impress. We can see the same approach in this book which is bound to motivate even the die hard conservatives to try these tools for improving the quality of products and processes.

This book has filled a definite void in Indian quality management literature and I am sure business managers and quality professionals will take advantage of it to put quality in India in higher orbit.

Lieut General H. Lal
Director General FICCI Quality Forum and
Chairman National Quality Campaign Committee
Quality Council of India

PREFACE

It is being recognized more than ever before that the *Quality is the Sharpest Weapon* to survive, compete and prosper. It calls for eternal endeavour of alternating sequences of activities for *control* and *break through.* This virtually implies imbibing *Quality Culture* by *all and sundry* in the organization.

Efforts need to be focused on analysis of *true, adequate* and *relevant data* to be efficiently effective. This in turn requires scientific bent of mind, may we call it a *Statistical Culture.*

Statistics is a science in search of *truth.* It is a science that *serves* all other sciences and is *master* of *none.* It is a *key technology* consisting of preparing a *problem bank, prioritising* these, collecting relevant right amount of data, analyzing these and making recommendations, *implementing* these, stabilizing *control* and *repeating* the cycle to achieve the goal of being a *leader,* through *continuous improvement* both *evolutionary* and *revolutionary.*

It is most unfortunate that *statistics* is, by and large, considered a very *dull, dry* and *complex* subject, so much so that it is shunned. An humble attempt has been made in this small volume of 141 pages to introduce seven tools that are *simple, elementary, easy to learn and practice quickly; so diverse in application and yet very highly cost effective with very vast potential for abundant returns.* These are all *fascinating.* The emphasis has been placed on their *concepts* and *applications* through illustrative *case examples* with a pious hope that it will serve the need of developing *self sufficiency* in the *use* of *statistical methods* for *continuous improvement,* at all levels in all fields of activities. An attempt has been made to bring home the truth that the *fruits of applying statistics are the sweetest.* There is no *escape.* Therefore *endear* it, the *sooner* the better, for the *prosperity* of the *society. Hopefully, it should generate interest beyond these tools.*

Magnificent seven refer to the seven tools namely *Cause and Effect Diagram, Check Sheet, Pareto Analysis, Stratification, Scatter Diagram, Histogram and Run Chart.* These are treated in chapters 2 to 8. These are preceded by a chapter describing the associated terminologies, concepts & economic significance and succeeded by chapters on a composite case study using all the seven tools, Organising for quality control and tips to imbibe Quality As a way of life through its ABC to habitually steer on the path of *Continuous Improvement.* The last chapter provides for Aptitude Test to assess Assimilation and the gap that needs to be bridged. These Elementary Basic Tools applied creatively embrace bulk of the problems, often faced in day to day work.

This booklet is the cumulative result of the dedicated and committed inputs received from the respected parents, teachers, literature, colleagues and the students in the class room and at work places consisting of several customer organizations, as also all members of my family. A select bibliography of authors and books is given at the end. I record my grateful thanks to all of them including human resource of associate organizations.

I request the readers to contribute to further enhancing its value by providing useful feedback after reading the contents and putting these into practice and earn my gratitude as well as that of the readers of subsequent revised versions.

A N Nankana
Patron, Quality Improvement Mission, New Delhi

CONTENTS

1

FUNDAMENTALS OF QUALITY

1.1 WHAT IS QUALITY?

Quality has been universally acclaimed as fitness for use, signifying the supremacy of the user in judging its adequacy. For instance it is for the user to decide how good a product is. Only the wearer knows where the shoe pinches. Likewise the user can tell how good is a pencil, a mixer grinder or a gear. Further, whether he is satisfied with the electric supply, its voltage including its stability and continuity.

- It refers to all types and sizes of products and services and in fact any activity. Product hence forth shall include service also.
- It is a composite result of a satisfactory design and its conformance.
- Conformance to specification(s) of product parameter(s) is (are) indirect approximate means of forecasting adequacy of product to satisfy the intended use. Product parameters include chemical, durability, mechanical, physical and allied properties. Like appearance, availability, brightness, colour (shade), composition, cost, diameter, ease of maintenance and replacement, fastness of colour, hardness, odour, purity, safety, timely delivery at desired place. This list is not exhaustive.
- Quality therefore encompasses all features like :

Appearance : that influences first sale.

Functional : that causes repeat sales and

Reliability : including availability and maintainability that sustains the market to ensure growth for survival. Lack of it is fatal sooner than later. Availability implies that the gadget works when operated or switched on while maintainability implies that repair, as and when it becomes necessary, is easy, economical and quick.

- Formal definition by International Organisation for Standardisation (ISO) read as:

Totality of features and characteristics of a product or service that bear on its ability to satisfy stated and Implied needs.

It now reads as

Degree to which a set of inherent characteristics fulfils requirements.

The preceding discussions only vindicate the above statement.

1.2 WHY QUALITY?

Consider mixer/grinder, for instance. A housewife shall be tempted to buy it, if it looks nice. She buys one, takes it home and operates it. It may not function as expected. She tells others about her experience. This message spreads like a wild fire and puts brake on its further sale. On the contrary, if she is satisfied, sharing her experience with others, will promote repeat sales. Further, as and when the service becomes necessary, it is reliable and effective life cycle cost is competitive, the organization not only sustains the market but is also able to ensure healthy growth rate. This generates more employment opportunities and hence prosperity for the society.

Quality indeed makes a difference between success and failure. Thus quality is necessary for survival, though survival is not mandatory.

1.3 WHAT ARE THE MEASURES OF QUALITY?

Some of the key *indices* of *quality achievement* are:

- Degree of *satisfaction* as reported by the user, Market complaints and Returns.
- *Productivity* Indices, namely Ratio of Conforming Output to each Input.
  Inputs include both direct and indirect. For example, energy, equipment, human resource, raw materials, and the *invisible yet instantly perishable* time. Each index needs to be monitored appropriately.
- Ratio of *Ideal* versus *Actual* costs of production .
  The Ideal cost of production is estimated under the assumption of *Zero Wastage* of inputs from conception of product to its fruition by *doing every thing Right First Time.*
- Measurement of Quality is the *Price of Non-conformance.*
  Account for every thing that need not have been done or would have been avoided if every thing were done the right way in the first instance and consider that as the Price of Non-conformance.
  Therefore, negatively, it is measured by the people with vision, as the loss that is caused to the society by the lack of quality or imperfection in the product delivered or service rendered.
  Imagine a Power House generating power lower than the designed capacity. The social loss is not just the loss of revenue to the producer of power, it also includes the loss of opportunity to provide better service to the existing users, and possibly meeting extra demand domestic or industrial. The latter adds to more employment potential providing necessary thrust to enhance Gross National Product and welfare of the society.
- Quality may also be simply expressed in terms of *percent non-conforming units or nonconformities*. Further if the criticality of various types of non-conformities is not uniform viz; the likely potential loss or damage caused is not equal or vary considerably, then an appropriate weighted demerit score is used as an index to signify the level of quality achieved. Inspection error, classifying conforming as non-conforming and vice versa, should not exceed one tenth of Acceptable Quality Level. Ideally the quality is zero non-conformity.
- Quality of an item with respect to any one measurable parameter may be assessed by its *proximity to standard or target* namely specified mean (middle value of the maximum and minimum specified acceptable limits called tolerance limits).

- Quality of a lot is best defined by the *pattern of proximity of deviations* of the observations *from the target* designated as zero. It should conform to the normal pattern or any other expected pattern depending upon the parameter being assessed and not violate the specified tolerance band. The process mean and standard deviation (root mean square deviation) together describe the quality. The smaller the variance (square of standard deviation) the better the quality. Ideally it ought to be zero. Instead Dr. G Taguchi's loss function, quantified as proportional to square of deviation, expresses quality in terms of monetary loss. This successfully highlights quality imperfections even though the product may be conforming to the specified norms. This might have been the single most important factor in focusing the attention of the top management on quality as hitting the target rather than satisfying the specification. This itch probably made Japan the world leader. It is obvious that loss arrived at through *loss function* is under estimate in comparison to the visualized *social loss*.
- Consider the game of hitting the bull's eye among four players A, B, C, and D. Compare the following four emerging situations. See Figure F1.1.
(A) is hitting away from the target and the hits are spread over a wide area.
(B) is hitting the target but the hits are spread over wide area like that in (A).
(C) is off the mark like (A) but the spread is over narrower area. He has better capability than (A). His failures are due to being off the mark. His performance will be better than that of (B) if only he can correct his bias, which is considered easier than improving one's intrinsic capability.
(D) is hitting the target and simultaneously in closer range too like (C). The performance Quality of hitting the target is best in this case.

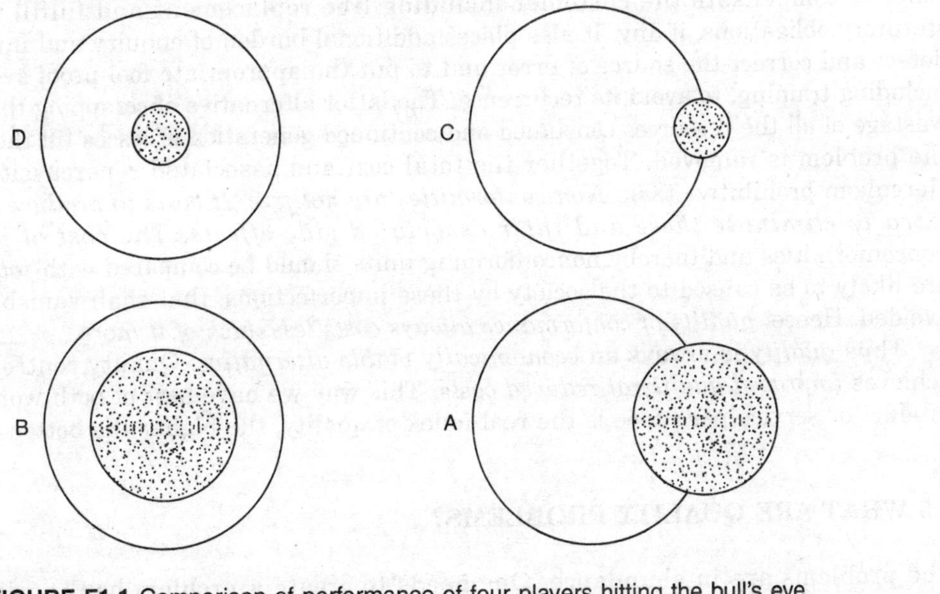

FIGURE F1.1 Comparison of performance of four players hitting the bull's eye.

In *defence* parlance, if one *fails* to hit the target , it survives to blow the *fatal* hit in return.

Thus quality is measured positively by the value, the user attaches to what in his perception he is receiving in return for the rupee (s) spent by him or negatively by the loss that lack of quality is likely to impart to the society.

It may also be expressed by percent non-conforming units or non-conformities or by appropriate demerit score or by the proximity of the value of the parameter of interest to the target together with its spread around it.

1.4 DOES BETTER QUALITY COST MORE?

It is often believed that *superior quality of design* costs more because of necessary costlier inputs. *This is not necessarily true.* Let us consider a practical example of two well known brands of scooters in India, Vespa (Bajaj) and Lambretta (Vijay super). Both these are designed to cater to the specific travel needs of particular economic group or market segment. Yet, one of these decisively established its overall superiority in design at lower cost. More examples are possible from housing designs to satisfy the needs of the customer in all respects at lower cost. *Thus it is possible to have better design at lower cost.*

Like wise, it is a common belief that production of conforming product entails stoppage of process for correction of errors that reduces production and productivity and enhances cost per unit of conforming product. Thus suggesting that quality of conformance also costs more. *This too, is false.* Consider the consequences of having produced a non-conformity and hence a non-conforming item. It has to be either reworked or scrapped. The former requires re-handling and re-inspection, increases inventory of in-process goods and re-processing. Re-processed goods are often not as good as the originals. If, however a non-conforming unit slips the inspection system and reaches the customer, it adversely affects the market share, generates fire-fighting and might lead to even struggle for survival besides involving extra costs to compensate the customer including free replacement and fulfill the legal and statutory obligations, if any. It also places additional burden of enquiry and investigations to detect and correct the source of error and to put the appropriate fool-proof system in place, including training, to avoid its recurrence. The latter alternative of scrapping the item implies wastage of all the resources consumed and continued generation of losses till the snag causing the problem is removed. Together the total cost and associated repercussions constitute Herculean prohibitive task. *Non-conformities are not free. It costs to produce these. It costs extra to eliminate these and their associated side effects.* The *cost of prevention* of nonconformities and thereby nonconforming units should be compared with *total losses,* that are likely to be caused to the society by these imperfections, that shall vanish if these were avoided. Hence, *quality of conformance always costs less, lack of it more.*

Thus *quality* is always an *economically viable alternative.* Quality control aims at and achieves *improved quality at reduced costs.* This way we have best of both worlds. Utility of product or service per rupee is the real index of quality, the higher the better.

1.5 WHAT ARE QUALITY PROBLEMS?

The problems are in abundance. One needs to create a problem bank, prioritise these, organize teams with proper facilitation and empowerment to resolve these as per planned schedule.

The quality problems fall in two distinct categories. These are *sporadic* and *chronic*.

Let us look at Figure F1.2. This shows sequence of production on X- axis and quality index on Y- axis, say percent non-conformities, the lower the better. The non-conformities in fifth batch are too low and in thirteenth batch too high to be ignored. Such occurrences here and there arise from what are termed SPORADIC problems. These have roots in lack of process control. The process is suspended and reasons for the abrupt change identified. These need to be inculcated in the former case, if economical and avoided in the latter case to benefit in either situation. These measures are parts of process control. Its organization include the following steps:

Assess Process Capability (Variation due to Chance Causes only)
Fix Optimal Target
Choose Appropriate Chart
Detect Abnormal Deviation
Investigate the Cause
Restore the status

Identify Uncontrollable Assignable Factors. Monitor these and Manipulate, such among the Rest Appropriately that counter the harm-full effect adequately. Reference is invited to IS: 397 parts 0, 1, 2, 3 and 4.

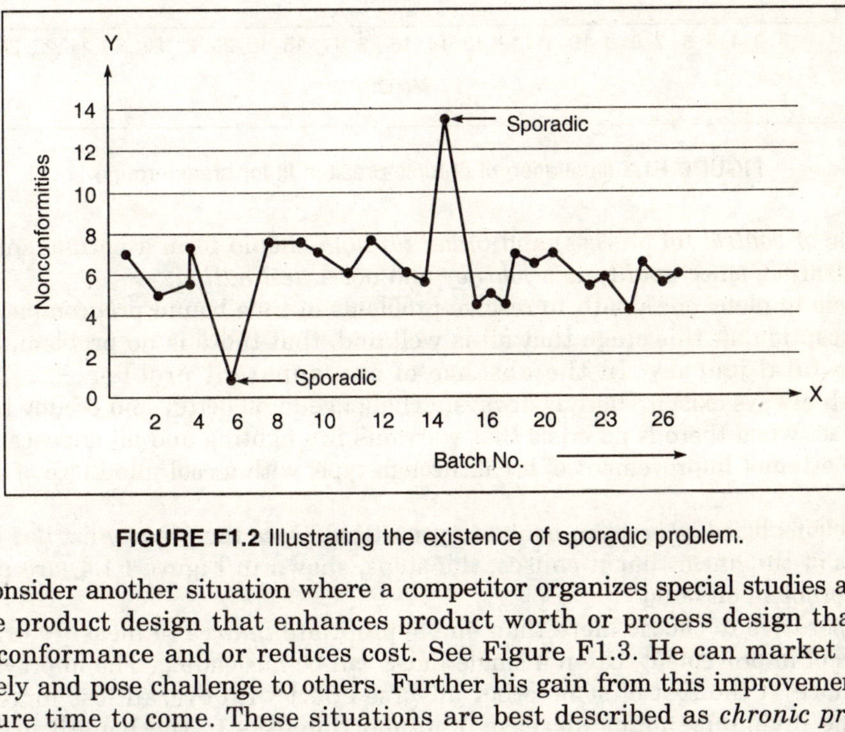

FIGURE F1.2 Illustrating the existence of sporadic problem.

Now consider another situation where a competitor organizes special studies and succeeds to improve product design that enhances product worth or process design that enhances degree of conformance and or reduces cost. See Figure F1.3. He can market his product competitively and pose challenge to others. Further his gain from this improvement is eternal for all future time to come. These situations are best described as *chronic problems* and approach to achieve this better level of performance is called *break through*. The steps to execute this programme include the following:

Developing positive change in the attitude
Laying down priority
Developing and executing relevant training programmes
Constituting Steering and Diagnostic Arms
Bringing in desired Cultural change
Organising Switchover to put the improved system in place

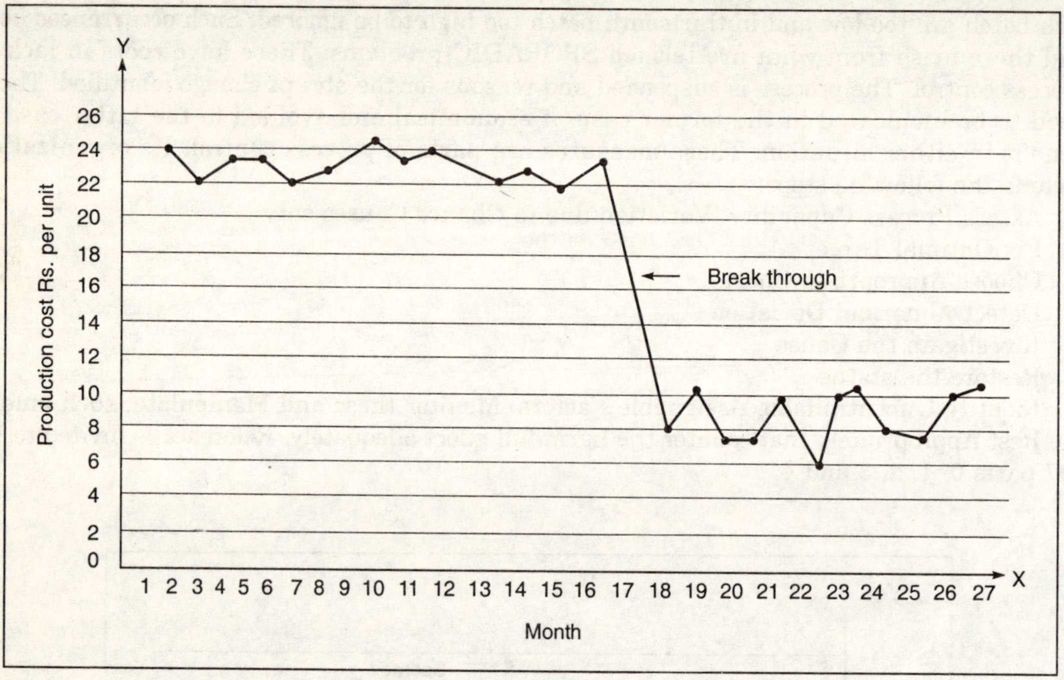

FIGURE F1.3 Illustration of chronic problem fit for breakthrough.

The cycle of *control* (of process) and *break through* should form a permanent feature of every organization, since *quality* is a *journey* and not a *destination*.

One needs to plans one's path to resolve problems in time bound programme.

Any perception, at this stage that all is well and, that there is no problem, signals the start of the fatal journey. In the absence of any apparent problem, the problem of breakthrough always exists. There is always a challenge to do better and occupy the space at the top. In fact when there is no crisis that warrants fire fighting and all is peaceful, it is the best time to attempt improvement of breakthrough type, with a cool mind free of any tension whatsoever.

The problems listed in problem bank are prioritized. Once the priority list has been made, on the basis of the harm that it causes; the steps, shown in Figure F1.4, are proposed to resolve the problem on hand.

It is imperative to choose more than one appropriate *indices* to measure or assess the effectiveness of improvement. Often a single index can be misleading. The improvement may be taking place at the cost of some other adverse effect with overall loss increasing. For example, the inventory might increase disproportionately to the benefit accrued from enhanced availability of the material. One may begin with a couple of indices.

Having decided on the indices, one needs to go in for planning the generation of *data* that will help valid calculation of the indices and provide link with the causative factors.

All the tools, that form the subject matter of this booklet, learnt should be innovatively applied to *analyse* the data gathered to derive maximum information and chalk out a comprehensive plan of action for improvement.

Take *action*(s) and *review* the improvement attained against the projected one's.

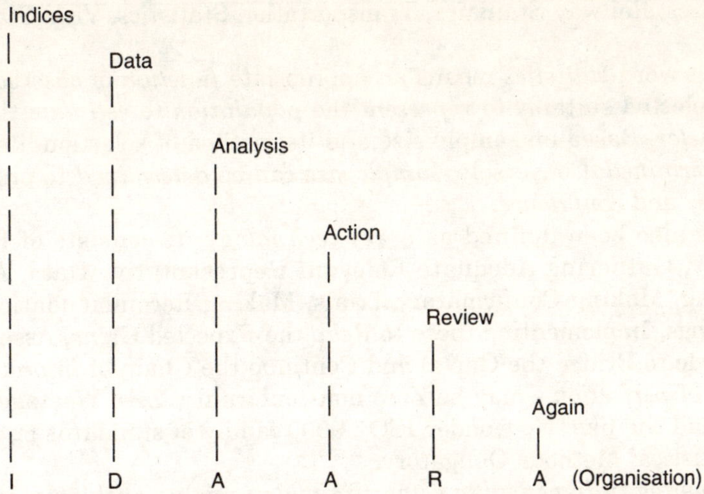

FIGURE F1.4 Steps to resolve the problem.
(The *urdu* word IDAARA means an organization)

The gap constitutes the basis for next iteration to repeat the cycle of activities for resolution of the problem for further *improvement*.

May it need be reminded, that, any glaring deficiency observed that can cause an imperfection or contribute to the problem on hand, even though unintended, need to have been satisfactorily attended to, before resorting to any structured approach to resolving the problem.

The culture of facing a problem as it arises is synonym with fire fighting. One successfully resolves the problem perhaps very speedily too, but then the organization and the society pays a very heavy price. The fire fighter might earn his promotion too! On the contrary, a system of prevention needs to be practiced as a way of life. This culture shall avoid the menace of firefighting. The system consists of anticipating the problem, developing a fool proof system to forestall it and to put the system in place.

1.6 WHAT IS THE ROLE OF STATISTICS?

Before attempting answer to this question, it is only fair to understand, what is statistics?

It is a science in search of truth and truth alone, nothing else but truth. It is a great pity that yet one of the quotes is; lie, damn lie, white lie and statistics. Statistics never tells lies. Some statisticians might do in the circumstances they are placed. Then telling lies is not their monopoly. It is not being said in their defence, but again a statement of facts. No doubt, the profession demands highest order of integrity.

Statistics is a science that serves all other sciences and is master of none. It delves into the void, *to know the unknown and traverses from uncertainty to certainty*. The un-intended risks arising from sampling and non-sampling *errors are contained within acceptable* norms with due consideration of long term *economic impact*.

The word *statistics* like several others has multiple meanings. Commonly it means any *data*. For example, Agricultural Statistics, Business Statistics, Commercial Statistics, Defence Statistics, Economic Statistics, Educational Statistics, Health Statistics, Intelligence Statistics,

Population Statistics, Railway Statistics, Transportation Statistics, Vital Statistics and so on and on.

Specifically the word *statistics* means an appropriate *function of observations made on a sample* of units selected suitably to *represent* the *population* to *estimate* the corresponding *population parameter*. Based on sample size and its method of selection, its *confidence level* can always be *determined*. Conversely, *sample size* can be *determined* to provide an *estimate* of desired *accuracy* and *confidence*.

Statistics has also been defined as a *key technology*. It consists of Formulating the Problem Precisely, Gathering Adequate Relevant Representative Data, Aptly Analysing, Validly Concluding, Making Confirmatory Trials, Making Recommendations Based on the Cumulative Findings, Implementing these to Reap the Expected Gains, Assessing the Gap(s), Repeating the Cycle to Bridge the Gap(s) and Continue the Chain of *improvements* to Reach the Evasive Goal of *perfection*—may be Zero non-conformity, Zero Wastage, Zero Deviation from the Target and the like. No wonder ISO : 9000 family of standards on Quality Systems makes use of Statistical Methods Obligatory.

Statistics provides indispensable scientific tools to solve problems of quality control including break through to sustain continuous improvement. The Basic tools of Statistical Quality Control include Cause and Effect Diagram, Check Sheet, Pareto Analysis, Stratification, Scatter Diagram, Histogram and Run Chart. These are Simple, Easy and Quick to learn and practice fruitfully. Reference is invited to IS : 15431. There are a host of other techniques to develop optimal solutions to problems encountered in almost all variety of situations.

Statistical methods are available to assess inspection inaccuracies arising from sampling and non-sampling errors. For the cumulative effect to be harmless, the error ought to be less than one tenth of acceptable quality levels and in no case in excess of one sixth. Generally these errors are found to be on the higher side. Statistical studies have helped to, reduce these to acceptable norms, standardize, and sustain these. These studies fulfill the obligations under the title Repeatability and Reproducibility Errors pertaining to ISO : 9000 family. This activity should precede planning for Process Control System and putting it in place. Statistical aids for process control have been mentioned in preceding section (1.5)

Statistical approach of experimentation for determination of optimal product and process designs have been exploited massively by advanced countries and in a limited way by developing one's. These help in getting valid conclusions with minimum of effort and investment. The optimal product parameters may constitute International and National Standards while optimal process parameters may form by and large only Company or Plant Standards.

As stated above, there are a large variety of other Statistical tools to cater to variety of other situations. These include decision to choose the product and location of the plant, choice of equipment; choice of suppliers, their rating and development; production scheduling, inventory and store maintenance, scheduling despatches, assessment of customer satisfaction and what not.

The above unambiguously makes it crystal clear that any worthwhile Total Integrated and Synchronized Quality Management Programme for Continuous Improvement *is incomplete* without adequate dose of *statistics*.

Adherence to systems for conformance, to optimal developed standard, assures quality and yield delivered at right time and place at economically viable and competitive prices.

Lastly, it needs to be emphasized that there is one and only one unique optimal or right quality for desired use to cater to specific market segment. A car or a motor bike run at unique optimal speed consumes least fuel, is more safe and comfortable on the road; minimises maintenance, service costs & pollution and simultaneously maximizes life of vehicle. This way it benefits both direct and indirect stake holders.

Quality level designed to be better than the optimal might cost more and render it uncompetitive. While quality level worse than the optimal will not do. In either case, it is not possible to sustain the market and the fall begins sooner than later leading to dissolution, the obvious logical end. Thus quality conforming to the rightly designed products and services benefits all stake holders, direct and indirect, the society at large. It requires thorough planning, execution, review and updating. This cycle needs to be sustained eternally.

Secret of Japan to become global leader in quality lay in harnessing its mass media network and institutional infrastructure to educate its entire human resource in the concepts, methodology and practice of simple quick and cost effective techniques. This enabled them to fully exploit the resources available to generate other necessary resources to create superior products that endeared the people world over. Japan took two decades to achieve this status. India has the potential to lead the world in much shorter time now. It only needs to start working earnestly.

2

CAUSE AND EFFECT DIAGRAM

2.1 WHAT IS IT?

Let us first list, all parameters or characteristics of a product, component, service or an activity of interest, which influence the satisfaction of the customer in particular and the society at large. Each one of these output parameters depends upon parameters of input materials, processes, equipments and allied resources including jigs and fixtures, environment, skill of workers, measuring instruments and test facilities etc. Listing of all these inputs called causes and its linkages with the outcome of concern such as yield, cost, non-conformity or excessive deviation called *effect,* when presented *diagramatically* is called *cause and effect diagram*. Each effect has a cause and conversely each cause an effect, nothing happens by itself. The law of *karma* is also the law of *cause and effect*. It is rooted in the universal law of as we sow, so shall we reap.

2.2 HOW TO DRAW IT?

Call a meeting of associated personnel representing Product Design, Process Design, Production, Inspection, Packing, Marketing and associated indirectly with the problem on hand or the effect of concern. From among this group a leader is chosen consensusly. He conducts the *brain storming session* by introducing the problem and inviting suggestion for the likely reasons or sources thereof. He presents these in the form of a diagram on the display board and keeps on updating it. The final outcome of several sittings and attempts may look like the one's shown in Figures F2.1 and F2.2. While so doing, care needs to be taken to avoid any action that might discourage the participating members to express their views frankly.

Conducting brain storming session is an art. The minds of the participants needs to be opened up. The leader has first to open up himself and endear others. A leader once had to ask the participants to forget about the problem on hand temporarily and instead enumerate the *various ways in which a glass tumbler could be used*. The participants gradually opened up and contributed to the *list of thirty six uses,* against a normal enumerated number of *5 to 10 ways*. The problem was taken up again. The participants were found to contribute more effectively.

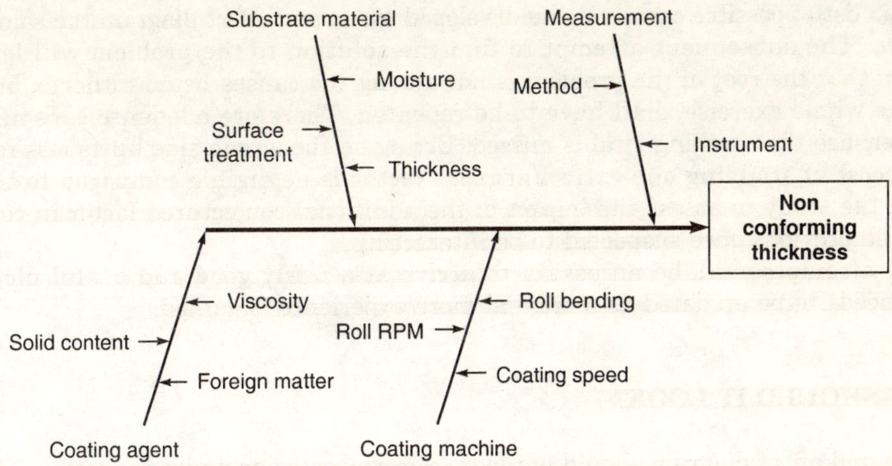

FIGURE F2.1 Cause and Effect diagram of nonconforming thickness of coating.

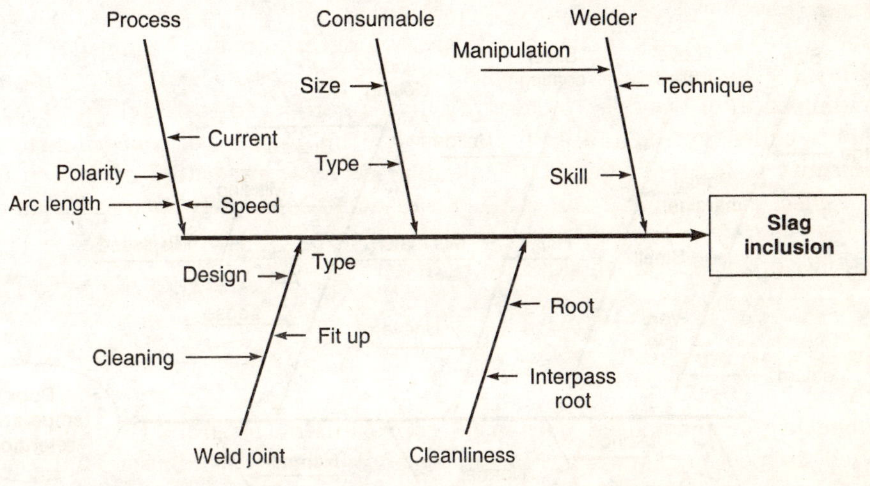

FIGURE F2.2 Cause and Effect diagram of slag inclusion in a cast iron product.

The attempt is to collectively *fix the problem and not the blame*. This is possible by thinking for *what went wrong* rather than *who went wrong. Who* went wrong or in *whose* presence the error crept in, is *vital,* since *he alone* is in a position to tell *how* all it happened and possibly *how* it could have been *avoided* or *how* it can be *prevented* in future.

Therefore a congenial atmosphere needs to be created where the team members are encouraged to share their experience in completing the diagram adequately. It needs to be exhaustive enough to include the *culprit* cause. Subsequently, it is these causes which shall be investigated for their contribution to the problem. If the culprit is not in the suspected or conjectured list of possible causes or the developed cause and effect diagram the solution shall be evasive. The subsequent attempt to find the solution to the problem will lead to the conclusion that the root of the problem is not among the causes listed hitherto, but beyond these. The whole exercise shall have to be repeated. Therefore adequate care needs to be taken to ensure that nothing vital is missed. Erring on the wrong side hurts less in the long run. The cost of studying one extra harmless factor is negligible compared to the cost of repeating the study to assess the impact of the additional conjectured factor in conjunction with all the previous ones suspected to be interacting.

A few attempts shall be necessary to arrive at a fairly good and useful picture. The diagram needs to be up dated as more and more experience is gained.

2.3 HOW SHOULD IT LOOK?

The cause and effect diagram should not look over congested or under congested. The former pin points that problem is complex or large and needs to be split. See diagram F2.3 for nonconforming thermos flask that failed to retain the desired temperature for the stipulated duration during final testing.

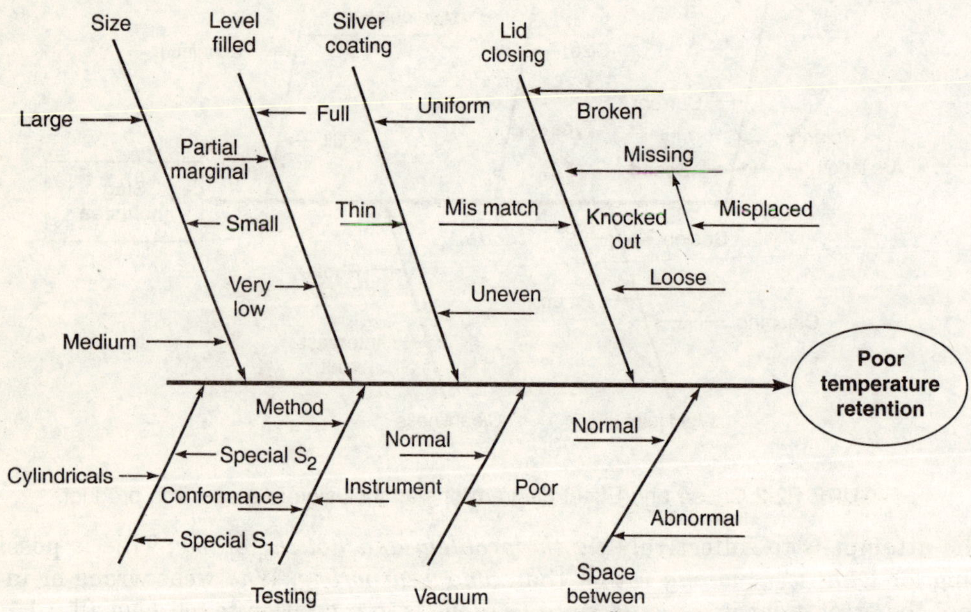

FIGURE F2.3 Cause and effect diagram of a thermos flask failing to pass the temperature retention test.

Each one of these sources can be split into specific cause and effect diagram. The one for silver coating is shown in Figure F2.4

The latter is indicative of not enough thought having been given to the problem needing another brain storming session probably by augmenting the team too, save exceptions. As an

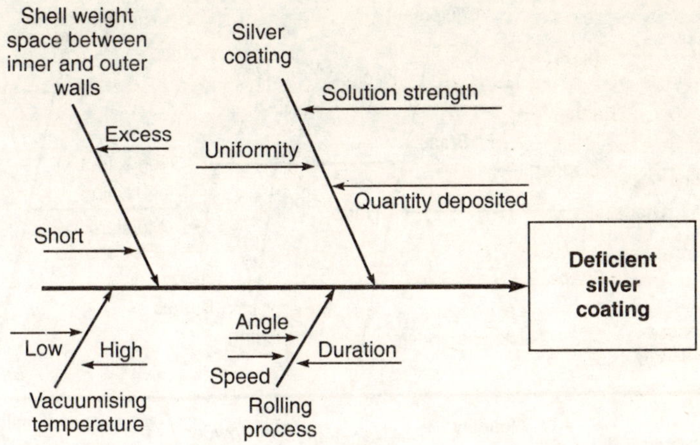

FIGURE F2.4 Cause and effect diagram of deficient silver coating, causing poor temperature retention in a thermos flask, both with respect to its quality and consumption.

example see Figure F2.5, showing cause and effect diagram for preparation of tasty tea. It is not adequate. The more complete version is shown in Figure F2.6.

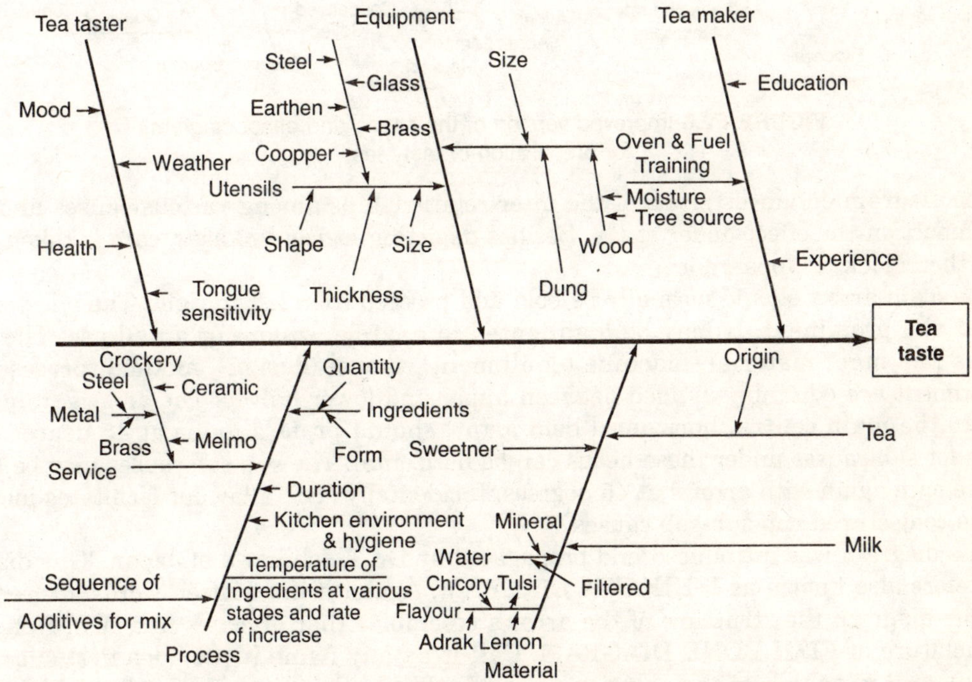

FIGURE F2.5 Cause and effect diagram for preparation of tasty tea.

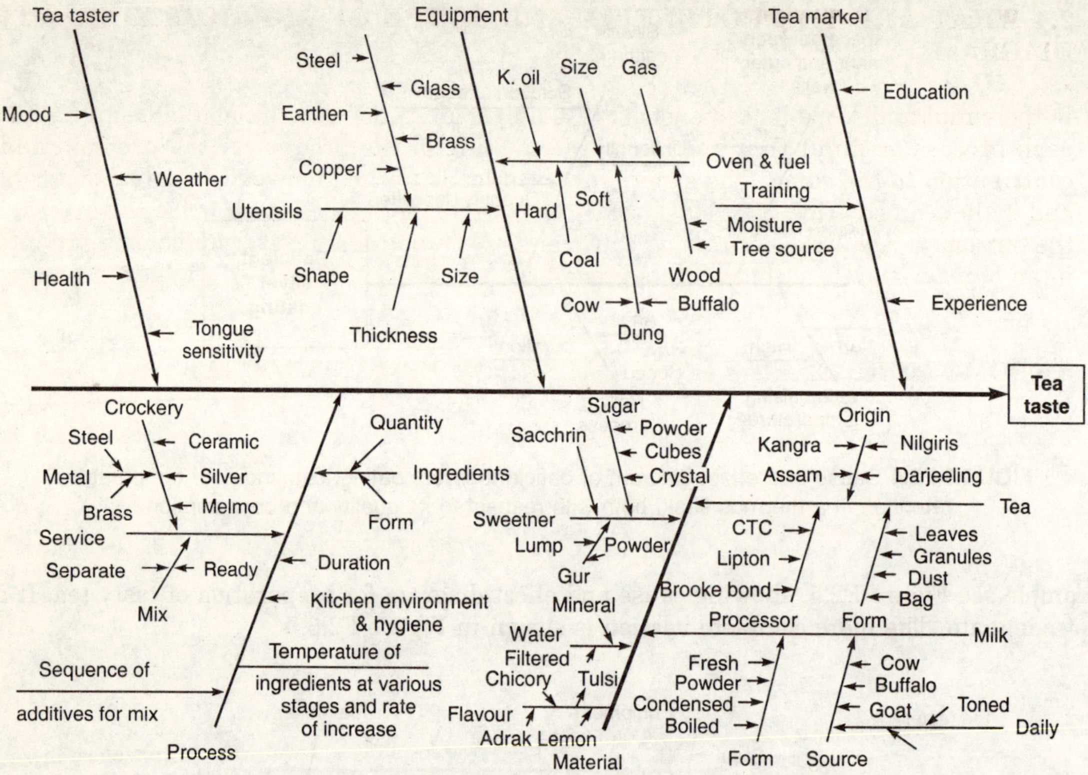

FIGURE F2.6 Improved version of the cause and effect diagram for preparation of tasty tea.

The diagram enhances grasp of the inter relationships among various causes and their total impact on the effect under study. Besides depicting logical linkages, care is taken about its aesthetic look or appearance.

The main arrow should normally be bold and proceed from left to right. The effect should be bold and prominent. It may be highlighted in a circle, square or an ellipse. The main sources like man, material, machine (equipment), model (design), method (process) and environment are equitably divided between upper and lower halves. The arrows from these titles to the main central horizontal bold arrow should preferably be at 45 degrees. The arrows for sub-causes under these heads can be horizontal. The sub-sub-causes may be linked to these once again with arrows at 45 degrees. Incidentally such a lay out facilitates inclusion of more conjectured sub-sub-sub causes.

This diagram was introduced and propagated by Dr. K Ishikawa of Japan. This diagram is therefore also known as ISHIKAWA DIAGRAM. Again, if the written words are removed from the diagram the structure of the arrows resembles that of bones in a fish. Hence the nomenclature of FISH BONE DIAGRAM. Call it by any name it provides first successful rational step in search of the solution to the problem. *One needs to get into the habit of starting the solution to the problem with drawing its cause and effect diagram.*

2.4 WHAT ARE THE POTENTIAL AND SCOPE OF APPLICATION OF THIS DIAGRAM?

Its potential and scope is tremendous. It is universal. It develops a habit of deep thinking of each process or input that goes in to the making of the product or the problem and its contribution to the same. The experience even leads to the knowledge of what factor hurts and by how much. This knowledge contributes to further efforts to optimize the process and the outcomes. A culture of drawing cause and effect diagram for every problem encountered in an organization is an index of its growth potential.

2.5 SOME MORE EXAMPLES OF CAUSE AND EFFECT DIAGRAM

Figures F2.7 and F2.8 depict the factors influencing the outcomes of performance of a Team in a game of Sport and National Status on Quality in the World Market, respectively.

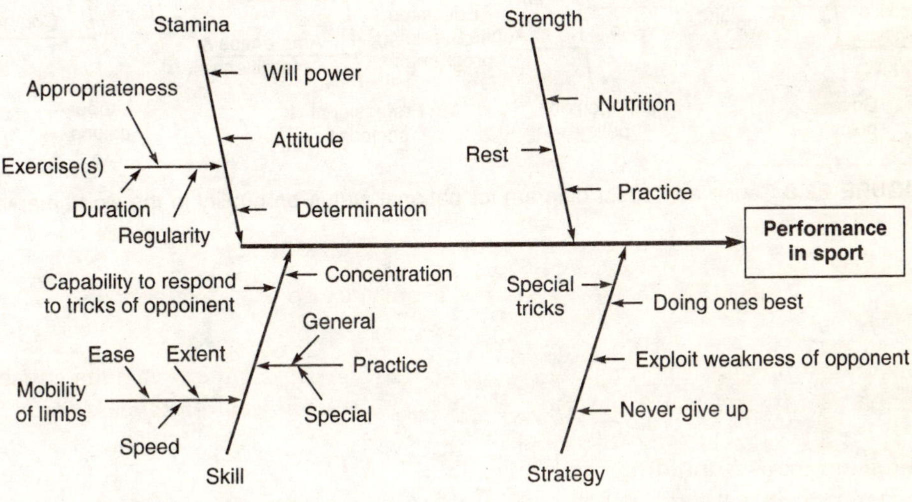

FIGURE F2.7 Cause and effect diagram of performance of a team in a game of sport.

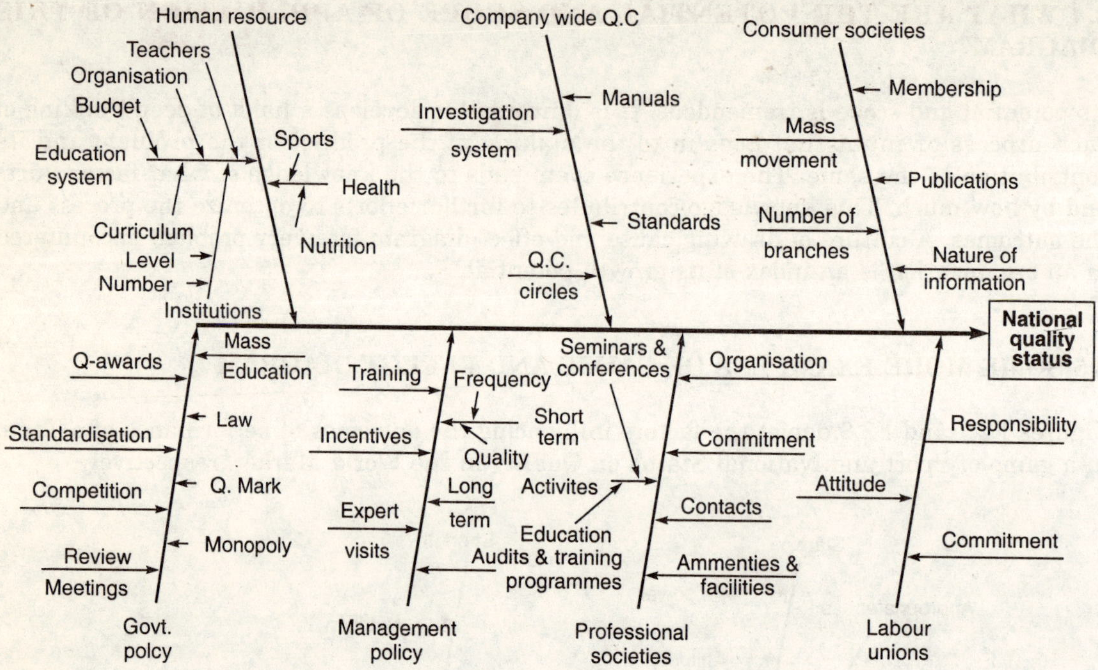

FIGURE F2.8 Cause and effect diagram for national status on quality in the world market.

3

CHECK SHEET

3.1 NEED FOR CHECK SHEET

A good quality control programme aims primarily at prevention of non-conformities. It requires generation of data on output parameter(s) and corresponding input parameter(s). The data thus generated are used to assess, whether what is being done is as required or planned. If however inspite of conformance to stipulated input parameter(s), the results are unsatisfactory, the data are analysed to determine new optimal level(s). Trial runs are made to confirm the projections. These are then implemented if found satisfactory. Otherwise the studies are repeated.

An appropriate standard format is necessary to record the information necessary to fulfill the contemplated needs for exercising measures for prevention and improvement. This format goes by the name of a check sheet.

3.2 WHAT IS A CHECK SHEET?

The format for recording all the necessary information to execute preventive plans and to enable development of strategies for improvement, commonly known as Data Sheet, Log Book, Inspection Record, Schedule of Enquiry has been termed *check sheet* in this context. Such a check sheet needs to be standardized to enhance its utility. It is better to arrive at its design in consultation with the users—who are to record the information, summarise and infer for immediate use or action and review for in depth analysis for feedback for improvement of control and to aid break through studies.

The nomenclature of check sheet has possibly emerged from the standard check list provided to the packer to verify and ensure that all the accessories meant for an equipment have been accounted for in the intended package for dispatch to the customer or the tick mark resorted to by the accountants to confirm that the item thus marked during the audit undertaken by them has been checked and found to be conforming to the stipulated rules and procedures.

As experience is gained the check sheets need to be continuously (periodically) upgraded and modified for improvements to serve the intended purpose more efficiently.

3.3 CONTENTS OF A CHECK SHEET

The contents of check sheet should be adequate enough to fulfill the contemplated needs of prevention and improvement. The following is illustrative list of contents.

- *Identity* particulars to enable *traceability* like organization, date, project, customer, supplier, specifications, batch or lot number, as applicable.
- Associated human resource and Equipments if any
- Input(s), implying consumption of resources of all kinds with pertinent details of their quantity and allied parameters like purity or strength or size as the case may be.
- Process parameters
- Environmental parameters of concern if any
- Method of sampling, sample size, frequency of sampling
- Inspection and test facilities used or availed
- Inspection and test method, least count or visual standards adapted
- Corresponding output(s) with details of parameters of interest
- Indices or appropriate functions of observations made

As already stated it is an illustrative list only and by no means exhaustive to cater to the diverse needs of huge variety of situations. Whatever is not considered worthy of the effort may be ignored and any other pertinent information missed needs to be included.

3.4 SALIENT FEATURES OF THE CHECK SHEET

It should contain all the vital information. It is a common belief that often unnecessary details are asked for. A standard question will address these doubts. Will it or will it not be useful in investigations, that may become necessary in the event of a nonconformity having been observed internally or reported by the recipient, to identify the cause or source of non-conformity and find a fool-proof solution to the problem ? If this is likely to aid, it needs to be included or else ignored.

It should be easy to document the required information and the necessary facilities to do so should be available. The amount of writing involved is minimised. The layout should facilitate summarisation, allied calculations and inference for associated decision making and follow up actions.

The cost of developing and maintaining a standard check sheet should be viewed from the likely losses, the lack of it might entail.

3.5 SOME ILLUSTRATIVE CHECK SHEETS

3.5.1 Attribute Inspection

A heavy electrical equipment manufacturer had a system in place to inspect, the fabricated jobs accomplished by the production to their satisfaction, by an inspector belonging to an independent inspection department. An inspection report of a huge fabricated job read as:

The job accomplished is shabby. The locations of non-conformities have been marked 'X'. The job may be re-submitted for inspection after necessary compliance.

The production department re-submitted the job after doing the needful. The inspector recorded the following observations in his next report:

It is false. The job has not been rectified in its entirety. The job needs to be re-submitted after completing necessary rectifications.

Has the inspector communicated his findings adequately to enable the rectifier to perform his role satisfactorily for a huge job of this magnitude, perhaps exceeding ten cubic meters with a large number of sections and segments? Locating all unspecified number of X's itself constitutes an uphill task. It would certainly have been helpful to give the following additional information(s) in that order:

- the number of X's marked on the job and their locations | to enable the task performer to do the job without omissions.

The locations may be descriptive like, front middle box inside top and so on, or indicated by an 'X' mark on the attached pictorial drawing.

Certainly, any additional information on:

- the nature of non-conformity
- the degree of its criticality indicating its likely potential harm
- the likely source of the fault, of course as opined by the inspector and
- the possible remedy (not documented in the work instructions in vogue) as a measure of suggestion

| may be given as a foot note or remarks

shall go a long way in planning and execution of the preventive measures.

The lay out of the check sheet plays an important role in making a summary to enable prioritization and strategic planning for necessary follow up actions.

A possible check sheet in this context is illustrated in Table T3.1. This is fairly elaborate and caters to a large number of similar situations. It is a three dimensional matrix. It can be

TYPICAL CHECK SHEET-ATTRIBUTE

Organisation: 'O' Department: 'D' Component: 'C' Drawing No: 'DN'
Date: Shift: Operator: Inspector:

Tally marks for nonconformities (defects)

Defect code	Criticality code	Location code				Total
		L_1	L_2	L_3	--Etc.	
	C_1	I				I
D_1	C_2					
	C_3	III				III
D_2	Etc.	—				
Etc.						
Total						

List of defect, critically and location codes should be provided.

N.B. 1. In simple situations location and or critical code may be redundant.
2. This check sheet is very appropriate for casting, fabrication, forging, ceramics and similar jobs.
3. Instead the job itself or the, 'drawings' of the component and its various sections or views may be used to indicate the location of the defect where the 'defect code' and its incidence can be indicated.

Table T3.1 An illustrative check sheet for an attribute data.

curtailed to address the needs of simpler situations adequately. In its present form, all that the observer is required to do to record his observation, is to put a tally mark in the appropriate 'cell'. Row totals, sub-totals of criticality categories, column totals, sub-totals for locations and grand total provide sufficient summary information for a job. Such respective totals over jobs enable identification of dominant sources to plan and execute steps for their avoidance and subsequent control to sustain the improved status.

3.5.2 Variable Inspection

An electronic company engaged in manufacture of audio equipments purchases lots of tiny components of large sizes (numbers). Its acceptance sampling plans chosen according to IS: 2500 call for samples of size fifty. The acceptance criteria and procedure require breaking of the sample into ten sub-samples of size five each; calculating sample ranges and means as also grand mean; calculating further 'statistics' for comparison with corresponding acceptable limits to ultimately decide on its worthiness for acceptance or otherwise for conformance to desired target and spread. Instead, check sheet shown in Table T3.2 simplifies data recording. It straight away builds the Histogram (See Chapter 7). Its interpretation with superimposed tolerance limits provides clarity about:

- approximate capability of the process
- process status in respect of central tendency and pattern of spread
- acceptability or otherwise of the lot, as also
- remedial measures required on the process to make it acceptable.

TYPICAL CHECK SHEET-VARIABLE

(Identity and traceability particulars)

Dimension class interval		Tally marks	Total
30.025	-30.025	𝍸 II	7
.075	.125		0
.125	.175		
-----	-----	----- LSL	
.175	.225	II	2
.225	.275	IIII	4
.275	.325	𝍸 𝍸	10
.325	.375	𝍸 𝍸 𝍸 𝍸 𝍸 𝍸 𝍸 𝍸 I	41
.375	.425	𝍸 𝍸 𝍸 𝍸 𝍸	25
.425	.475	𝍸 𝍸 𝍸 𝍸 III	23
.475	.525	𝍸 𝍸 𝍸 𝍸 𝍸 𝍸 𝍸 IIII	39
-----	-----	----- USL	
.525	.575	𝍸 I	6
.575	.625		0
.625	.675	𝍸 III	8
		Total	165

N.B This is very useful for
1. Incoming inspection
2. To confirm setting target when it is fast moving job but likely to last for short duration.

Table T3.2 showing illustrative check sheet for variable inspection.

For example the data of Table T3.2 indicates that:

- the process is not centered at the desired mean of 30.325 and instead is bimodal meaning that the process is mix up of produce of two distinct averages viz 30.350 and 30.500.
- there are sporadic incidences of a few avoidable over size and under size components
- the process seems to be capable of producing components conforming to desired tolerances.

3.5.3 Packing Process

Machinery manufacturers often receive complaints that the spares accompanying the machine were incomplete. They even put in a claim for the replacement and the damages caused by the delay in the completion of the project. The remedy lies in preparing a standard check sheet (list) for each equipment. The packer then ticks each item in the list as it is packed and thus avoids complaints and avoids additional costs necessitated by extra correspondence, packing and dispatch besides compensating for the damages claimed. A possible specimen of suitable check sheet for such situations is shown in Table T3.3.

TYPICAL CHECK SHEET
(Identity and traceability particulars)

List of items and quantity to be packed and or despatched to a customer (say as spares)

S. No.	Item	Size	Quantity
1.	Fastener		
2.	Screw driver		
3.	Gasket		
4.	Weight		
5.	Rubber washer		
6.	Keys		
	Etc.		

N.B. Similar lists may be made for performing and or inspecting certain jobs or activities such as

1. Assembly of cars, tractors, turbines etc.
2. Maintenance of buildings, machinery etc.
3. Count down for space program with due regards to their sequence.

Table T3.3 A typical check sheet for list of items to be packed and or dispatched to a customer such as a standard list of spares accompanying a machine tool as accessories.

3.5.4 Work Sampling Study

Work sampling study consists of observing the equipments or even human resource at random (un-predictive) intervals of time; on whether they are busy, if so on what activity and consequences of interest; or if idle, the source or reason there of. Such studies are aimed at augmenting the efficiency of the system. The proposed check sheet for such purposes is illustrated in Table T3.4.

CHECK SHEET FOR WORK SAMPLING STUDY

Company: 'C' Department: 'D' Period: 'P' Superviser: 'S'

Operator Number ↓	Machine or round number → 1	2 (Time)	3	4	Etc
1					
2					
3					
.					
.					
Etc					

Entries in codes

Working status if working:		Cause of nonworking status:	
No feed, idle running	WI	No material	NM
Workng at half capacity	WH	No orders/plan	NP
Non conforming material	WNM	Electrical breakdown-waiting	NEW
Non conforming process	WNP	Mechanical breakdown-waiting	NMW
		Maintenance electrical in progress	NEP
		Etc	

Table T3.4 Illustrative check sheet for conducting work sampling study.

3.5.5 Consumer Complaints

One of the most important driving force for under taking projects for improvement by special designated teams are the dissatisfied customers. Their feedback is crucial. Apart from what the customer may have to say on the subject, he needs to be guided to provide additional information, that is vital for resolving the problem or incorporating the improvements visualized by him, to enable focus on the specific aspects for necessary development. A sketch

TYPICAL CHECK SHEET
(Identity and traceability particulars)
Consumer complaints for equipment

Town	Usage type	Month/year of manufacture	Nature of failure code A	B	C	Etc.	Total	Suggestion.

N.B. 1. Town describes environment conditions

Table T3.5 A specimen check sheet for recording Feed back from Customers.

of check sheet to cater to this need is illustrated in Table T3.5. It needs to be modified to meet the demands of specific situations.

3.56 Setting Approval

The prerequisite for process control is to start right. The settings for the process parameters ought to be perfect or as much perfect as possible. The record of settings made will be very handy and useful for effective control of the process through use of appropriate process control chart. A specimen of a possible check sheet for approval of setting prior to commencement of production is shown in Table T3.6.

CHECK SHEET FOR SETTING APPROVAL

Organisation : Machine No: Product:

Characteristic: Specification: USL.....LSL.....TOL......

Setting approval: (USL + LSL)/2 ± TOL/4

Serial No.	Date	Time	Observation 1	2	3	4	Total	Mean	Deviation from target	Remarks	Initials
1	2	3	4	5	6	7	8	9	10	11	12

Table T3.6 A specimen check sheet for *setting approval.*

3.6 HOW MANY CHECK SHEETS TO PLAN?

The Cause and Effect Diagram shall provide adequate guidance for stages needing the use of check sheet(s), its type and the details of its contents. As mentioned earlier, the check sheets need to be revised and updated in the light of experience gained and more so to fulfill the requirements of the relevant updated cause and effect diagram.

4

PARETO ANALYSIS

4.1 WHAT IS PARETO ANALYSIS?

Alfred Pareto an Italian economist, discovered during his research studies, that it is a law of nature that wealth (income) is concentrated among the vital few. About twenty percent of the of the top few (the vital few) possessed about eighty percent of the total national wealth while the remaining majority of eighty percent were contented with the rest of the twenty percent wealth. This versatile law of nature is called Pareto law. This is so true, that twenty percent of the over 500 countries the world over share among themselves about 80 percent of the total global natural resources while the rest eighty percent have to manage with the remaining twenty percent innovatively. Like wise, of the about 28 states and union territories in India, only 5 to 7 account for the major share of 80 percent among themselves, while the rest share the remaining twenty percent. Even in sports, take for example cricket, it is only a few 2 or 3 out of 11 member team who make most of the runs or claim major share of wickets. Even the office efficiency depends more on the committed about 20 percent human resource, on whom the boss depends for timely and correct delivery. This law, jolly well, extends to the distribution of non-conformities or the loss emanating from these, among its many possible causes or resources. The art of analyzing the data to identify the vital few sources of losses emanating from lack of quality, is called Pareto Analysis. For illustration see Figures F4.1.

4.2 WHAT IS ITS USE?

To maximize the result of efforts and allied inputs to resolve quality problems, it is worthwhile to concentrate on the vital few sources for their avoidance or avoiding their effect, instead of dissipating the limited energy on the trivial many. This is synonym with the policy of divide and rule, except that there is no need to perform the unethical role of dividing. The non-conformities by law of nature are already so rooted among its causes or sources. It is we, the human resource, who need to be united against the inanimate non-conformity to, not only eliminate it but also to prevent it from recurring or taking its birth again.

Lest, the message is misunderstood, it needs to be reemphasized that we must always work in unison as one cohesive team, if only we wish to succeed in our mission, whatever it may be. Divide and rule policy may be fine in the short run but certainly disastrous in the long run. We are welcome to do so, if we are selfish enough to care for ourselves and neglect our next generations. We therefore prosper at the cost of those whom we love most and for whose sake we believe or pretend to be working. Let us be honest to ourselves.

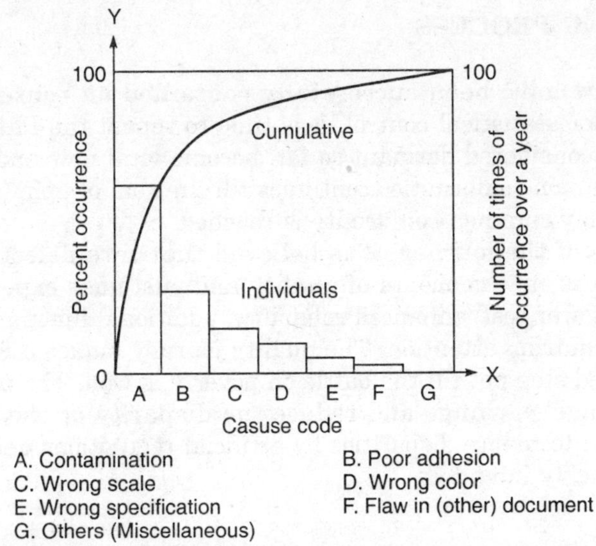

FIGURE F4.1 Pareto Analysis. Concentration of nonconformities by types of problem (cause).

4.3 HOW TO DO IT?

Once adequate and appropriate information has been generated in the check sheet(s), the non-conformities in the end product or service confronting us are classified according to their causes or sources, since these emanate from the operational levels of the parameters which quantify the respective inputs that are to be acted upon selectively.

The causes are then ranked in diminishing (non-ascending) order of their frequency of incidence or contribution to the loss, if the harm from the different causes is uniform. The findings are then pictorially presented in the form of a bar chart, with the 'ranked' causes on the X-axis and the 'percent contribution' on the Y-axis. The cumulative figures, cumulating to 100, are shown as a curve on the same diagram. Reference may be made to Figure F4.1, once again for its comprehensive look.

4.4 PRECAUTIONS

Care should be taken that the contribution from the category of 'Others' or 'Miscellaneous' causes is indeed negligible and preferably does not exceed 10 percent. If it is in excess, it may be split into the major among these and the rest.

On the contrary, if the 'number one', the most dominant is too dominant and too complex to be resolved, it should be suitably split or subdivided to get a hang over the manageable part of the problem. The small success will pave the way for enhanced confidence and capability for success in the next iteration.

It may be worthwhile to mention, at this stage, that the cause that ranked number two hitherto becomes number one for the next iteration and so the phenomena or the struggle for step by step continuous improvement continues.

4.5 A CONTINUING PROCESS

When the vital few have been successfully contained on routine basis, namely their contribution is under statistical control, it is time to repeat the Pareto Analysis study. The causes which were considered dormant so far, become vital now and need to be attended on priority. The war on non-conformities continues till the goal of ppm (parts per million) or ppb (parts per billion) nay zero non-conformity is reached.

If at any stage of the journey, it is believed that zero defect has been reached, the competition brings in new concepts of quality and customer expectations; may be visual appeal, longer life, lower cost, enhanced reliability, additional functions that again show up as non-conformities requiring attention. The quality journey makes a fresh start.

Arise, awake and stop not till the battle on poverty is won. The prosperity shall be ours if we raise the 'average' earnings and reduce the disparity or the 'spread'. The short cut convenient approach to reduce disparities by artificial regulations defeats the very purpose of the noble mission nobly intended.

4.6 CONTRIBUTION TO LOSS IS PERTINENT

Some times it so happens that the loss contributed by a particular non-conformity is many times over than some other type of non-conformity. In such situations the Pareto Analysis should be based on proportional contribution to the Quality Loss by the non-conformity rather than on the basis of relative incidence or frequency.

4.7 SOME EXAMPLES

4.7.1 Chemical Plant

Pareto Analysis of concentration of non-conformities.

The summary of number of non-conforming batches of an adhesive for one month's produce is presented in Table T4.1.

Table T4.1. Showing Nature, Frequency and Rank of Non-conformity.
Organisation: ORG Product: PRO Month: MO/YR

Serial Number	*Non-Conformity*				
	Parameter	*Frequency*		*Rank*	
		Number	*Percent*	*Roman*	*Alphabet*
0	1	2	3	4	5
1.	Colour	31	50.0	I	A
2.	Composition	6	9.7	III	C
3.	Contamination	10	16.1	II	B
4.	Sticker	5	8.1	V	E
5.	Testing	6	9.7	IV	D
6.	Viscosity	4	6.4	VI	F
	Total	62	100.0		

It give details of Nature (Cause) of Non-conformity, Frequency (number and Percent) and Rank number with Alphabetical Code.

The above Table is recast in Table T4.2, after listing the nature of non-conformity in the order of their respective ranks with alphabetical designations.

Table T4.2. Showing Rank, Percent frequency and percent causes.
Organisation: ORG Product: PRO Month: MO/YR

				Non-conformity	
Serial Number	*Rank*	*Parameter*	*Cumulative percent of successive parameters*	*Percent Individual*	*Frequency Cumulative*
0	1	2	3	4	5
1.	A	Colour	16.7	50.0	50.0
2.	B	Contamination	33.3	16.1	66.1
3.	C	Composition	50.0	9.7	75.8
4.	D	Testing	66.7	9.7	85.5
5.	E	Sticker	83.3	8.1	93.6
6.	F	Viscosity	100.0	6.4	100.0

The plots of Percent Parameters or *causes* on X-axis (column 3) versus percent contribution, individual on Y-axis (column 4) in the form of a bar and cumulative on Y-axis (column 5) in the form of curve are shown in the diagrammatic form in Figure F4.2 known as Pareto Diagram. It shows that resolving the first cause will bring substantial gains. After resolving this, the Pareto analysis needs to be repeated to sustain attempts for continuous improvement. Possibly, the cause at rank two may acquire the status of rank one. The process is thus kept alive.

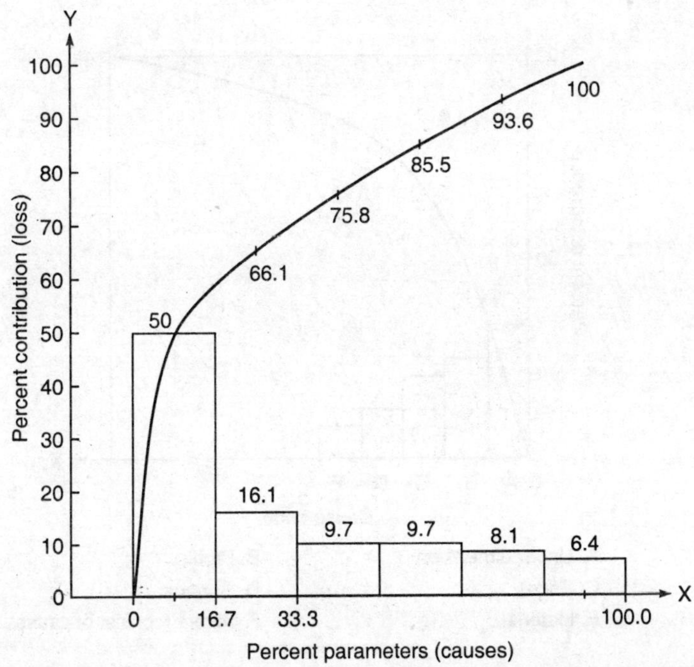

FIGURE F4.2 Pareto Diagram of nonconforming batches of an adhesive.

It is too obvious to state that, if the remedy for any non-conformity down the line is known, fool-proof system to avoid its recurrence needs to be put in place immediately rather than waiting for its turn indicated by its rank. Such exceptions only prove the rule.

4.7.2 Heavy Electrical Plant

Pareto Analysis of non-conformities detected in Turbine Blades.

The relevant data are shown in Table T4.3.

Table T4.3. Showing cause wise percent contribution to nonconformities among Turbine Blades.

Serial Number	*Cause code and description*	*Percent causes cumulative*	*Percent contribution to nonconformities*	
			Individual	*Cumulative*
0	1	2	3	4
1.	A Linear Dimension	10	32	32
2.	B. Profile	20	25	57
3.	C. Edges	30	19	76
4.	D. Surface	40	12	88
5.	E. Material	50	7	95
6-10.	F to J. Others (Misc.)	100	5	100

The plot of column 2 on x-axis versus column 3 on y-axis as histogram and column 2 on x-axis versus column 4 on y-axis as curve presents the Pareto analysis based on the respective individual and cumulative contribution from various causes. This is exhibited in Figure F4.3.

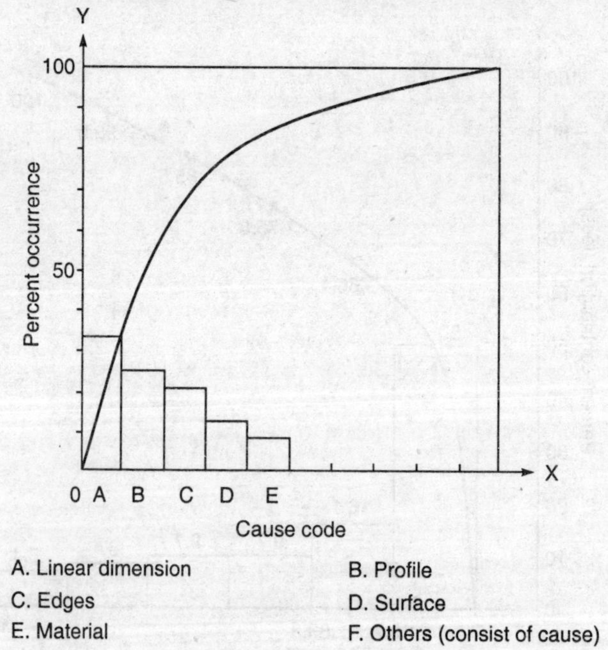

FIGURE F4.3 Pareto analysis of nonconformities amogn Turbine Blades.

4.7.3 Foundry

Figure F4.4 illustrates Pareto Analysis from a cast iron foundry.

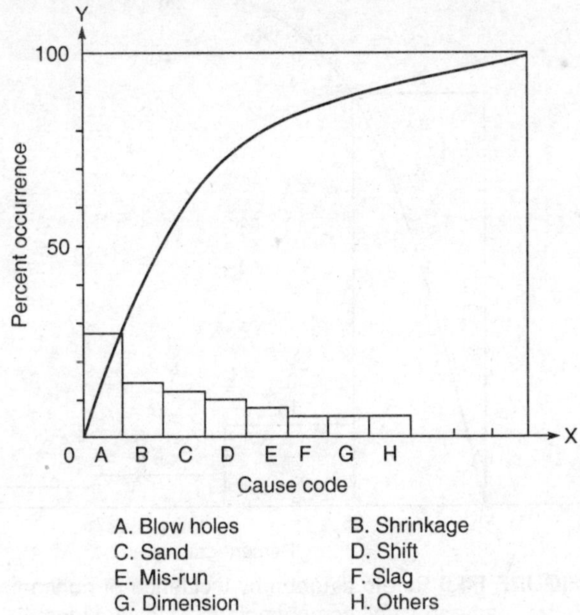

FIGURE F4.4 Pareto Analysis of nonconformities in a foundry product.

4.7.4 Engineering Industry

The details of non-conformities observed among small components in a heavy industry with respect to its frequency of occurrence are shown in Table T4.4.

Table T4.4 Data on nonconformities observed among small components in a Heavy Industry.

Serial Number	*Cause code and description*	*Percent nonconformities*	*Rank*	*Rank wise percent nonconformities*		
				Cause code	*Individual*	*Cumulative*
0	1	2	3	4	5	6
1.	A. Curvilinear dimension	15	II	B	75	75
2.	B. Linear dimension	75	I	A	15	90
3.	C. Surface	6	III	C	6	96
4.	D. Others	4	IV	D	4	100

The plot of column 6 along y-axis versus cumulative percent causes namely 25, 50, 75 and 100 along x-axis as shown in Figure F4.5 depicts the corresponding Pareto Diagram.

Additional information was later provided that each nonlinear nonconformity costs ten times as much to rectify as it costs to rectify a linear dimension. The cost weights for other nonconformities were also ascertained. This impacts the priorities for finding the solution to the sources of nonconformities. The revised summary is given in Table T4.5.

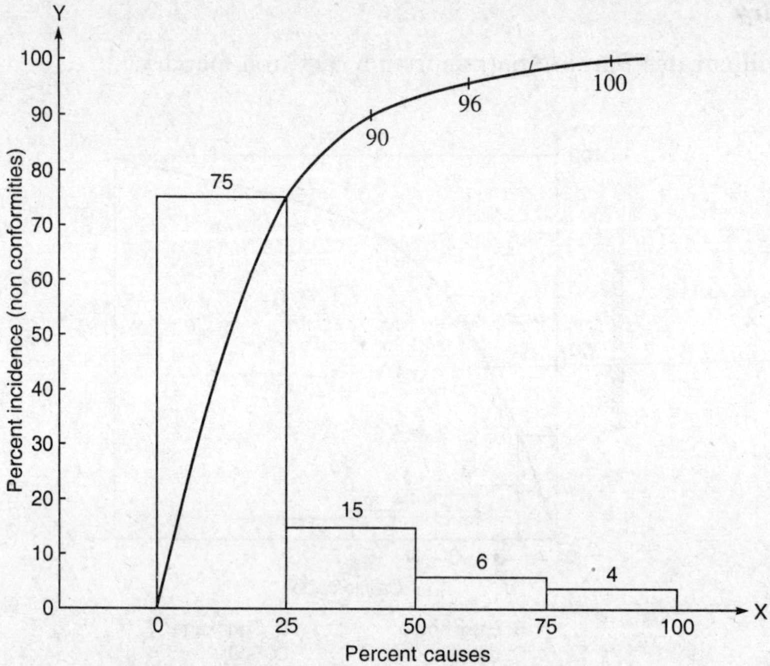

FIGURE F4.5 Pareto diagram by incidence of nonconformities among small components in a heavy industry.

Table T4.5. Data on nonconformities observed among small components in a heavy Industry by causes and their proportionate contribution to loss from rework.

Serial Number	*Cause code and description*	*Percent nonconformities*	*Loss weight*	*Loss (Relative) Amount*	*%*	*Rank*	*Cumulative loss %*
0	1	2	3	4	5	6	7
1.	A. Curvilinear dimension	15	40	600	65.5	I	65.5
2.	B. Linear dimension	75	4	300	32.8	II	98.3
3.	C. Surface	6	2	12	1.3	III	
4.	D. Others	4	1	4	0.4	IV	100.0
	Total			916	100.0		

Notes: Causes C and D together contribute less than 5 percent loss. It is therefore Desirable to club these.

The loss from cause A is too dominant. It may have many sources or subcauses. The relevant details are desirable to focus on manageable area in first iteration.

The revised version of the Pareto Diagram appears in Figure F4.6. This gives a picture more close to reality than the one already seen in Figure F4.5.

4.7.5 Service Sector—Telephone Exchange

The cause wise data on inoperative telephones are shown in T.4.6, both by the frequency of its incidence and the duration of its remaining nonfunctional. In each category these are ranked in non-ascending order of their contributions.

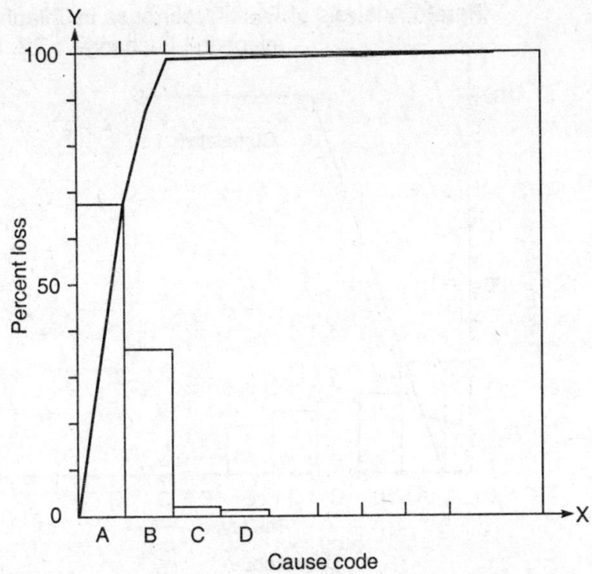

FIGURE F4.6 Pareto diagram by percent loss contribution by nonconformity types among small components in a heavy industry.

Table T4.6. Fault Code of Inoperative Telephone Instrument with frequency of its incidence and idle duration.

Telephone Exchange: TE **Period: PE**

	Incidence—Number of complaints				*Idle duration-hours*			
Rank	*Fault code*	*Number*	*Cumulative*	*Percent*	*Fault code*	*Hours*	*Cumulative*	*Percent*
I	A	1490	1490	30	C	7539	7539	32
II	B	1005	2495	50	D	5867	13406	57
III	C	954	3449	69	B	4452	17858	76
IV	D	883	4332	87	A	3261	21119	90
V	E	507	4839	97	E	1432	22551	96
VI	F	102	4941	99	F	458	23009	98
VII	G	51	4992	100	G	246	23255	99
VIII	H	9	5001	100	H	117	23372	100
IX	J	6	5007	100	J	93	23465	100

Fault Codes: A : NFF External; B : Cable; C : External; D : Subscriber's Apparatus; E : Subscriber's Fittings; F : NFF Exchange; G : MDF; H : Hot; J : Exchange

The respective Pareto diagrams are shown in Figures F4.7 and F4.8. The consolation is that either way the first four faults remain the same. Thus for customer satisfaction, the first four causes accounting for a major hurt to the customers and loss to the Telephone Department, need be addressed on priority.

It is pertinent to emphasise at this stage that the repeated use of the three tools, described hitherto, in this sequence helps to diminish the size of the problem. We need to iterate till we

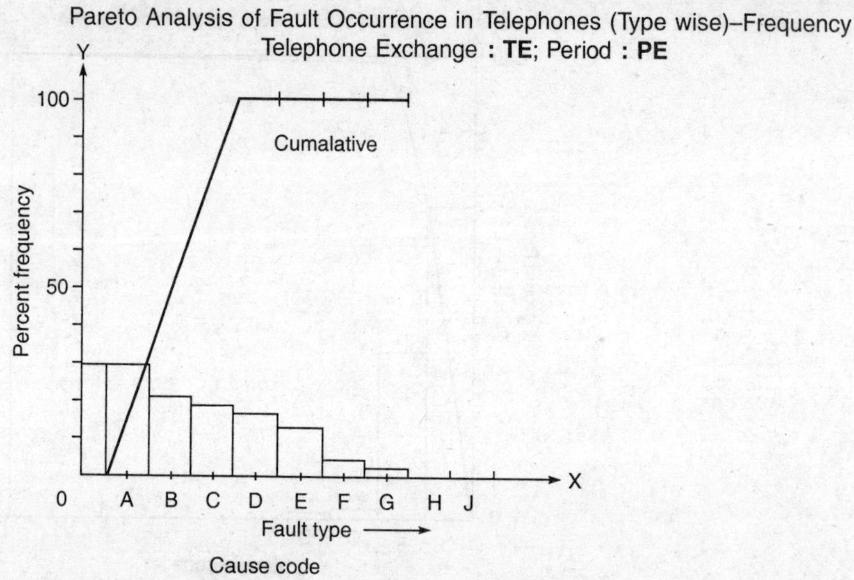

FIGURE F4.7 Pareto diagram of data of Table T4.6–Incidence of occurrence.

Fault Codes:

A : NFF External	B : Cable	C : External
D : Subscriber's Apparatus	E : Subscriber's Fittings	F : NFF Exchange
G : MDF	H : Hot	J : Exchange

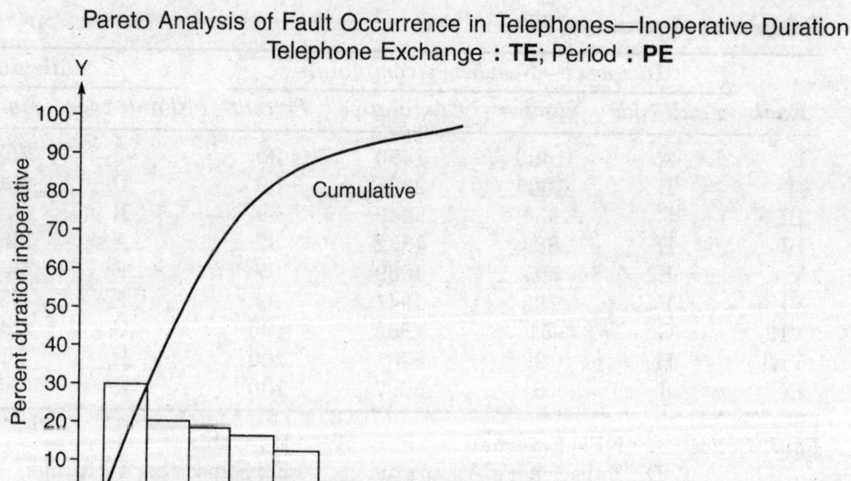

FIGURE F4.8 Pareto diagram of data of Table T4.6—Inoperative duration.

are in a position to crack the problem. To optimize the fruits of our efforts, we need to bite as much as we can chew. To begin well is a good strategy. Well begun is half done. Our first iteration deserves preferential consideration.

5

STRATIFICATION

5.1 DEFINITION

Stratum has been defined as a layer or a set of successive layers of any deposited substance. Strata is the plural form of stratum.

Stratification is the process of classifying the data into groups such that the groups are as much homogeneous as possible 'within' and heterogeneous 'between'. Thus inter groups or strata disparities or variation may be large but within or intra small.

5.2 UTILITY

This technique is used in socio-economic and allied surveys to obtain maximum efficiency in estimating appropriate indices. It aims at reducing sample size and simultaneously the error of the estimate. In this manner, it minimises the use of resources besides reducing the duration for completion of the task on hand. All these together contribute to substantial savings of the exchequer.

5.3 PRE-REQUISITES

In industrial and service sector applications, the pre-requisites for deriving maximum benefit from the application of this concept are:

- Identification of the quality problem through Pareto Analysis.
- Drawing of a Cause and Effect Diagram through 'Brain Storming Session' among personnel of all disciplines associated with product or activity directly or indirectly and allied stake holder; that is, listing or enumerating all factors that influence the quality parameter of product, service or activity under study; and
- Developing suitable check sheet for recording true data on quality parameters with traceability to the corresponding operational level of each factor including all pertinent inputs.

If above stipulated conditions are satisfied, then from among all the seven simple quick and cost effective techniques, the Stratification costs least and delivers most. In fact, if the data are not amenable to stratification by the levels of factors, conjectured to influence the effect, the parameter of the output of concern, the data are useless. What data to collect, how much to collect, to what accuracy and the manner of recording needs to be planned optimally, for mission to succeed timely and economically.

5.4 PROCEDURE, SALIENT FEATURES AND POTENTIAL

Once the data have been collected, the stratification approach implies, classification of each quality parameter by levels of one or more factors at a time depending on total amount of data and technological knowledge of their interdependence.

Quality is known to be influenced by design or model, process or method, equipment or machine including allied jigs & fixtures, location & environments, testing equipment including allied facilities and instrumentation, process parameters, human resource or worker, maintenance, sequence or time of production etc. The term 'Human resource or Worker' has been used in a very broad and general sense. He may take the form of operator, inspector, helper, packer, user or a team. Thus quality parameter of interest is classified according to the levels of each of the above relevant conjectured *factors* of influence, one or more, *if they are interdependent,* at a time.

5.5 SUCCESS STORIES

5.5.1 Stratification with Respect to Workers

(a) Vanaspati (Hydrogenated Oil) Production Unit

A vanaspati production unit was manufacturing its own 16.5 kg tins for packing, in its auxiliary unit. About 16 percent of the tins were found to be leaking during one hundred percent post production or pre-filling inspection, as a measure of control to avoid subsequent losses. Manufacturing operation consisted of stamping top and bottom square pieces; cutting sheet, shaping it to cover all the four sides and seaming the side to make a hollow cuboid; seaming the hollow cuboid with the top and the bottom, followed by soldering on the seamed side.

Examination of past one month's data revealed steady performance of rework of about 16 percent as stated earlier. It is therefore an apt problem of Break Through. A dialogue with the inspectors further revealed that the non-conformity of leak was localized and mostly detected on the side seam. This by itself, is an illustration of informal geographical stratification. This operation is operator dominant. It was therefore decided to study the failures operator wise.

Six workers were performing the soldering operation. The worker (soldering operator) wise classified summary disclosed that, the rework was almost 'nil' for 5 of them and almost the entire rework was emanating from the 6th one only. The examination of the 6th worker displayed the obvious. He had a deformed thumb. He was assigned another suitable job and a trained substitute worker remedied the situation. The savings in the form of resulting negligible inadvertent re-work, elimination of re-handling and enhanced production are worth the small effort made to remedy the situation.

(b) Tin container manufacturing unit

Another factory, producing tin containers of the same size as in (a) above, was supplier to the producers of hydrogenated oils. It was facing a peculiar problem of 'non-acceptance' of the lots by the vendee. A check inspection at the vendee's site confirmed prevalence of about 10 percent leaky containers. The chief of the manufacturing unit was amazed, since he had implicit faith in the competence and loyalty of his inspection team engaged in doing one hundred percent pre-shipment inspection check of the tins. He therefore, conjectured the possibility of damage in transit causing the leaks. If the conjecture of the chief, that the damage to the empty tins occurred during transit of short distance to the customer's site were true, the consequences of damage and resulting leakage loss and accompanying mess during transit of filled tins by the customer to his customers could be disastrous. A sample check, of tins ready for dispatch, confirmed prevalence of ten percent leaking tins and established that the stage of inspection was indeed the weak link in the system. A sample check of the output of each inspector was done. It revealed about 30 percent evading detection for three of the inspectors and practically 'nil' for the remaining seven. It so happened that these three workers were union leaders. They enjoyed the confidence of all. They were blissfully ignorant of the ground reality. Incidentally, they were giving more output and enjoying production bonus too. These workers became conscious and cautious, on knowing the problem and its repercussions.

Never the less, the output of the inspectors was stacked separately and samples were checked for confirming the 'zero leakage' status prior to dispatch. There was 'nil' complaint from the customer for next six months, thus eliminating the problem. This saved the business for the good of all!

Control of manufacturing process, including improvement if any called for, to produce zero non-conformity that will avoid the existing necessary evil of one hundred inspection plus sample check would be better alternative.

(c) Pharmaceutical Company

In one of the bottle filling and packing units of a pharmaceutical company, four workers were engaged in identical operation of labeling. Output of one of the workers was twice that of the rest of the individuals. The movement pattern of the hands of these workers was observed for a short while. The pattern of movement of their hands differed, leading to differential units of work or energy (ergon). The science dealing with such aspects is called 'ergonomics'. The best worker was promoted to train all the workers, not only in this unit but also in all other filling and packing units. He was empowered to reorganize these suitably to enhance the overall performance. Thus a new standard procedure was established and put in place for this kind of work. This approach paid rich dividends.

(d) Electrical Transformer Manufacturing unit

In a plant manufacturing oil transformers, it took a team of 4 to 8 workers to assemble a unit in 2 to 8 hours depending on the model. All transformers thus assembled are checked for being leak proof. The rectification process, in the event of a leak being detected, took couple of weeks. Thus the 'nuisance value' or tangible and intangible losses together caused by a transformer failing to pass the leakage test, are too obvious to be spelt out.

The average rework was about 22.5 percent. There were several teams or groups of workers performing similar tasks. Team wise stratification of their performance for past six months was accomplished. It was found that two of the teams delivered one hundred percent conforming (non leaky) transformers. Based on this part of the information, true to expectation, the performance of the worst group was as bad as forty five percent non-conforming (leaky) transformers.

Detailed discussions as a part of the investigations and attempts to find the reasons there of, in search of the solution to the problem, it became known that the leading pair of teams had achieved 'zero non-conformity' by assembling components thoughtfully selected from the available stock. Workers of the other teams assembled the left over components which lacked fitment criteria. They did not spell out this fact, for reasons best known to them. One of the reasons could be that there was no incentive for good performance by individuals nor any disincentive for poor performance.

The remedial measures rightly concentrated on development of supplier(s), standardization of packing, transportation, preservation and handling practices to ensure input of components of right quality for assembly.

(e) Export Inspection Agency

Export Inspection Agency Inspects goods on sample basis and certifies it export worthy, to prevent complaints from abroad and to build reputation for Indian products. Inspite of this, there are complaints. The inspectors get benefit of doubt on the basis of sampling errors as also on the factors presumably beyond their control after they have certified these to conform to stipulated standards. A two way table of 'sufficient' data classified by 'inspectors' and the recipient making a complaint revealed the true story, of course after providing for 'reasonable' benefit of doubt arising from sampling errors. It identified the dispatching and receiving stations lacking integrity.

To sum up, these are glaring examples of study teams getting at the root cause of problems to eliminate these. It simultaneously avoids mutual infighting on hunches, to blame each other for the survival of the individual. Alternatively, the existing culture of poor performance survives and havoc continues. Team work, instead, eliminates the problem for every one to survive.

Generalising, the meaningful stratification with respect to workers, helps in identifying the best, moderate and poor workers. The observation of a sample of workers from each of these groups generates the knowledge of DO's and DON'T's for performing the task on hand efficiently. The establishment of standard methods and training in their adoption yields many breakthrough results in a large variety of situations in all kinds and sizes of industries and service organizations.

5.5.2 Stratification with respect to machines

(a) Chemical–Rayon pulp production unit

A rayon grade pulp manufacturing unit had an elaborate system of input and output parameters at each processing stage, namely, chipping, digesting, bleaching and paper making. However, no purposeful analysis of data was done for any effective action for improving quality or augmenting production. The inconsistent quality of chips in respect of its size, in the chipping house, was adversely affecting the subsequent processes. It was safely attributed to heterogeneity of the quality of incoming bamboos from the forest, classified as dry, green, rotten or yellow.

A simple tailor made special study, called work sampling, was organized. Each chipping machine was observed at random intervals of time. The quality of bamboos (dry, green, rotten or yellow) being fed into the chute of the machine and the quality of chips (size) being produced were recorded. Supplementary information gathered consisted of the number of bamboos in the chute of the machine, the condition of the blades and its fitness. The examination of data thus obtained revealed that the quality of chips in respect of size, was both conforming and non-conforming irrespective of the quality of bamboos fed. Further investigation revealed that the cut quality of chips depended more on the sharpness of the chipping blades loaded on the machines and their adjustment. In depth scrutiny disclosed that sharpness angle of the blades and their adjustment on the machines influenced the interval for re-sharpening as also its life. These have direct bearing on the economy of chipping operation. It also highlighted the need to monitor the feed of the bamboos. It was seen to be very very erratic. Any shortage or excess in the number of bamboos being fed directly influence the productivity and life of the blades.

This is an illustration of rewards of stratification with respect to machines. It belied the myth that the fluctuations in the quality of chips with respect to their size were caused by the changes in the status of bamboos received for chipping. The myth prompted, no action. As a result, heterogeneous chips were fed to the digesting house. It created problems for the subsequent operations. A machine wise periodic check, of input and output, was introduced to make adjustment or replacement of blades, as a part and parcel of process control procedure.

(b) Electrical Engineering—Filling of cylindrical shells and sealing caps in assembly of capacitors

Two identical machines were in use for assembly of capacitors. Their day to day performance was satisfactory. The confidence gave place to complacency and hence process control procedures if any were being ignored. One day the customer's order could not be complied with. The production was short and the quality was poor. As usual a fire fighting investigation started.

It found that, one of the two machines was running smoothly. Its output was normal and the finish of the components was excellent. The output of the other machine was less. The finish of its component was poor. The component would often get jammed. The operator would be required to extricate the jammed component before restarting the machine. One could see a heap of crumpled shells (scrap). He was under pressure to produce to meet the demand for the day. The machine was in need of repair.

The pros and cons of doing the needful for the machine before commencing production versus running it for the day are too obvious to be taken lightly. The company paid a heavy price for giving a go by to the maintenance. The price included the cost of segregation, partial compliance of the order and more damage to the machine adding to the cost of inevitable maintenance and associated intangible loss of customers goodwill.

5.5.3 Stratification with Respect to Time

(a) Textile Mill

Generally, in a textile mill, the productivity during day shift is better than in the night shift. However, in one of the mills, the converse was found to be true. The investigations lead to the identification of healthy practice of planning production. All small and special jobs, likely to need attention from superior authorities, were deliberately executed during the day shift. This enabled timely aid, from the right quarters, on all matters needing attention. Hurdles, if any, in the path of production were taken care of. All the remaining hassle free jobs were planned for the night shift. This was the secret of enhanced performance during the night shift. So much so, that the total conforming output for all the three shifts was way ahead of other competitors. This even demystified the theory that when the cat is away mice play, implying that in the night shift the work is lax.

(b) Environmentally sensitive industries

Classification or stratification with respect to shift also brings out the effect of environmental temperature on the product, provided the job is not labour intensive. Alternatively, the differences are the combined effect of 'shift environment' and the skill & stamina of the worker performing the task. The contribution of each can be assessed by classifying the data simultaneously with respect to shift and worker or group of workers, as the case may be. In some, situations even the day of the week has its own influence. In such cases three way classification is called for. Often fresh data are not necessary. The past data suffices. Workers are routinely changed in shifts, often weekly. Past three weeks data are adequate. If necessary, it may be collected a fresh. Three way classification of data by three shifts, three weeks and three workers or teams, provide very useful assessment of the effects of respective factors being sought to resolve the problem. In addition the differences among the averages for the six days of the week provide the measure of the systematic differences among days, if any and sow the seeds for developing strategy for improvement.

(c) Engineering Industries

Stratification with respect to time clearly brings out the effect of wear and tear of tools to guide, the formulation and adoption of standards for intervals and magnitude of adjustments called for. This enables optimal use of tools and simultaneously assures production of desired quality right first time.

(d) Chemical Industries

Similarly in chemical industries, stratification aids in assessing the rate of fall in strength of chemicals like pickling and plating baths to enable formulation of optimal standards for timely additions and replenishment inclusive of their quantities for adoption. This also assures quality right first time. It makes optimal use of chemical and reduces rework to enhance efficiency.

Stratification with respect to time from the commencement of the job gives signals for any appreciable deterioration, particularly for time dependent activities or jobs, before it is too late.Timely action avoids subsequent hazards. A stitch in time saves nine.

5.5.4 Geographical Stratification

(a) Pharmaceuticals Marketing

A pharmaceutical company was satisfied with its growing sales and accompanying profits. The sales were stratified region wise. It revealed that growth rate was different for different regions. Not only that, in some cases, the sales had declined. Sales of each region was then classified by products to identify the dominant products contributing to this adverse situation. Together these constitute, two way classification. This helps in going into causes for the fall more specifically. May be, new substitute products have been introduced by the competitors or weather changes have influenced and diversified the demands or something else has happened. In either case, the discussions, with these findings as the basis, shall guide the approaches for alternative remedies to make up for the shortfalls in the sales of individual products in specific regions and enhance the total market share. In fact, the discussions lead to policy of aggressive sales in virgin areas.

(b) Shoe Manufacture

A large shoe manufacture company took a policy decision to prolong the life of the shoes manufactured and marketed by them. They introduced a scheme to offer concession to customers exchanging their old shoes for new. This ploy helped the company to collect a large number of old worn out shoes of various brands of their competitors besides their own. A fairly large representative sample from among these was examined for locations (geography of the shoes) that had worn out, rendering it unusable. This location wise stratification was used to identify the location prone to fastest wear ; the next location, the 2nd fastest and the likely gap between the two, and so on. This information was useful in developing alternative strategies. The final decision, on prolonging the life of the shoe depended on the estimated additional cost versus extra life possible and the resulting economy depending on the worth that the customer is likely to attach. What a strategy to win over the market?!

The concept is equally effectively applicable to component dominant assemblies like watches, automobiles and compressors; as also foundry and fabrication. Innovative use of 'geographical' concept is necessary to exploit its full potential.

(c) Instrumentation

The factories manufacturing auto meters or water meters and the like do one hundred percent inspection to identify the non-conforming meters. These are then rectified and rechecked before clearing for marketing. The stratification with respect to components requiring adjustment, repair or replacement indicates the specific components that require improvements. Each one of these is then taken up as a problem for solution by the respective teams. Many factories have made use of these strategies to more than double their conforming production with the same set up and resources in a short span of time with opportunities for several more such iterations.

(d) Electrical fan manufacture

A ceiling fan manufacturing unit was doing one hundred percent inspection for its conformance to noise, speed and wattage. The roof had provision for nine hanging positions for inspection. During inspection care was taken to use the same set of Blades, Condenser and Tachometer. Voltage too was stabilized. Classification of non conforming fans with respect to these positions revealed position biases in respect of speed. The nine positions in a square room were, as shown in the lay out in Figure F5.1.

Same fan at Position P5 will give maximum speed; at Positions P2, P4, P6 and P8 medium speed; and at the remaining four positions at the corners namely P1, P3, P7 and P9 the lowest speed. Thus the same fan had highest chance of acceptance at P5, lower at even positions and still lower at the remaining odd positions. This problem was resolved by assessing the biases and allowing corrections for the same. The results at position P5 were taken as the acceptable standard.

Differences among analysts and laboratories are not uncommon. These arise from non-standard practices and dispersions among testing facilities such as equipments or instruments including materials and chemicals used for testing the item. Such practices need to be standardized to satisfactory levels so that the total errors are harmless.

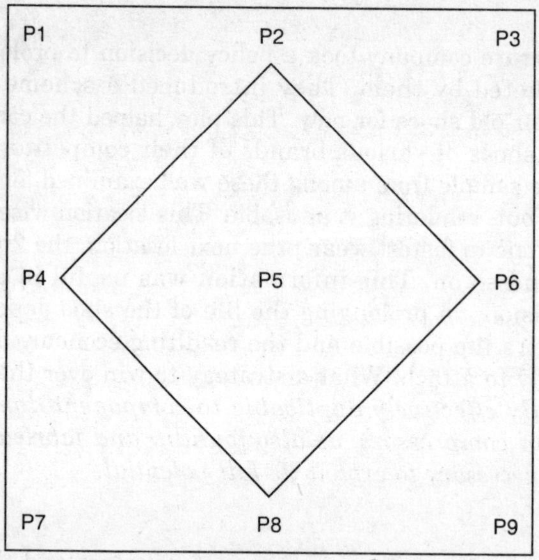

Figure F5.1 Showing layout of the fans for inspection on the ceiling.

(e) Service Sector—Food Grain Ware House

A national food grain ware housing body as usual engaged in procurement of food grains in areas that have surplus or abundant production and moving the same to the needy areas that are deficit. There were some losses in transit. The problem was to estimate the same and also identify the areas or routes that are more prone to such losses.

A sample study was planned to observe marked bags at selected dispatching and receiving stations (geographical locations). The weights dispatched and received were recorded. In addition a questionnaire was sent to these locations, seeking information on the weight dispatched by the sender and the weight received by the recipient. This information was summarized in a matrix, a two way table, illustrated in Table T5.1. This, as mentioned earlier, is known as two way classification.

Percent weight received. Excess (+), Short (–)

Recipient Station	*Despatch Station*					*Overall*
	A	*B*	*C*	*D*	*E*	
V						
W						
X						
Y						
Z						
Overall						

Table T5.1 Showing excess or short weight reported.

· By looking at the last column and the last row, one can comment on the integrity of the sender and the receiver in recording of the weights dispatched and received by them respectively, in addition to comparing the proneness of the routes with regard to losses in transit. The findings of the physical survey and the response to the questionnaire were in agreement. This added confidence in the conclusions drawn.

(f) Service Sector—Laundry attached to Hospitals and Hotels

The laundries practice standard procedures for washing linen received from different locations (wards or rooms, kitchens or banquets as the case may be). Depending on the capacity of the machines and the models, standard regime for quantity of detergent, temperature, speed and duration are followed. There are occasions when the washed lot is not considered sufficiently clean and either it is given extra run or re-washed. Thus the capacity is under utilized. Instead, the procedure should not mix up the linen from different locations. In other words, the linen is stratified by locations. Each stratum is then treated on its merit or needs. The standards in terms of quantity of detergent per kg of linen, temperature, duration and the like should be developed for each source depending upon the nature of spoilage. This did help the laundries attached to hospitals and hotels to improve their performance. It resulted in better customer satisfaction, increased life of the linen and reduced cost through improved utilisation of the installed capacity and allied resources.

(g) Fabrication, forging and foundry

In fabrication, forging and foundry operations, the two way classification by the nature of non-conformity and location is of immense value to the chemist and allied technocrats to pin point the likely few causes or sources of the non-conformity. These are then examined to identify the culprit for corrective measures. The leading industries are successfully exploiting this approach. Can others do without this?

(h) Pharmaceutical—Tablet Making

Homogeneity of the ingredients of the tablet and its weight are important parameters.

The former, even if achieved during its preparation, is likely to get distorted, particularly in *unani* medicines in powder form, during tablet making. The vibrations of the hopper, cause lighter ingredients to move up and the heavier to move down. Thus the ratio of the ingredients gets distorted and is not same for different layers or strata of the contents in the hopper. The remedy lies in avoiding over filling of the hopper, a practice often resorted by the operators as a matter of convenience. Depending on the nature of the mix, a standard height to which the hopper can be loaded needs to be developed and stipulated for conformance. If this is not done, even the weight of the tablets will deviate from the intended norm, and so will be the ratio of the ingredients. The tablets will not provide the standard dosage or the effect to the patient.

In case of multiple punch tablet making machine, the periodic samples of a few consecutive tablets are taken to verify that the weights and their dispersions conform to the prescribed norms. This procedure is good enough, if the assumption that all punches are delivering the same average weights, holds good. Unfortunately, often this is not so. In such cases, the data thus obtained, are likely to show more dispersion between tablets of the same sample. This in turn over estimates the process variation or its capability. It may show up the process as incapable, unless relatively the tolerances are wide enough. Alternatively, the process may appear to be in control, when in fact it is not. If this is the case, it is indicated, either by the runs of points (see chapter 8) on one or the other side of the standard line OR by the concentration of points close to the standard or the average line. The remedy in such situation consists of special check of each punch for any bias for over weight or under weight beyond acceptable norms. This should be periodically re-affirmed. The periodicity depends on the degree of the duration for which the calibration remains stable. Also, the sampling method needs alteration. A sample of a few consecutive tablets need to be taken from the same punch. This is punch wise stratification. This method of sampling is possible, only after the revolving table of the punches has completed the round or the cycle. The number of rounds shall equal the number of tablets in the sample. The data thus plotted on the chart is more likely to show up punch biases, assuming that the process has remained fairly stable between successive samples. Yet, another option could be to treat a set of all tablets from one round as one 'unit'. A sample of at least two units is called for. The sample statistic should be adjusted or corrected for the number of punches on the table.

5.5.5 Stratification with Respect to Supplier

(a) Vanaspati (Hydrogenated Oil) production unit

In spite of conforming to stipulated regime, the end product did not conform to the laid down standards. Production department blamed it on the laboratory responsible for testing and acceptance of the oil purchased, and the latter in turn blamed the former. The test records were being maintained supplier wise. The non-conforming batch of vanaspsti could not be traced to the supply because the oil received from different sources would get mixed up when transferring to huge storage tanks. All test records showed that only conforming good oil had been procured. Yet, the problem was there for all to see.

The higher authorities advised them to resolve the problem together instead of blaming each other. It was agreed to trace it to the storage tank with identity of the suppliers of its contents. A team was formed to over see the efforts and provide necessary support. The past record was criticality examined tanker wise, with details of the suppliers, whose supplies made up its contents. It took some time to lay finger on the supplier, who was common to the contents of the storage tanks, the oil from which resulted in non-conforming batches.

It was decided to keep a special eye on the future supplies of the suspicious supplier. As the next supply arrived, the seal of the tanker was checked. It was found to be intact. The sample of the oil, taken from the exit pipe as usual, was also found to be satisfactory on test. However, the doubt lingered on. The truck (tanker) operator was asked to break the seal and open the lock for inspection of the tank. He refused on one pretext or the other, such as, he had instructions that seal must not be disturbed in any manner whatsoever, he did not possess the key. He was over ruled. He was promised compensation for the loss caused by breaking the seal and the lock. What a surprise ! The tank was partitioned into two halves, interconnected at the bottom. The sample from the other half was tested. It was found to be highly adulterated. This in turn made the entire oil of the receiving tank, unfit for production process envisaged on the basis of the available test reports. Any production from this tanker would yield unsatisfactory result. Thus the remedy was known. The unscrupulous supplier was black listed, for this deliberate mischief.

The worth of maintaining records, that can enable tracing of the occurrence of non-conformity to its source, can be recognised, at least now. I S O : 9000 stipulates as one of its major requirements. The details need to be good enough for the purpose, neither too much nor too little. The former adds to the cost while the latter defeats the purpose.

(b) Electrical fan manufacturing unit

The entire range of components needed for assembling a complete fan are painted. The quality of paint is judged by the glossiness, surface hardness and whiteness it imparts as also the surface area covered per litre of the paint. The factory was maintaining satisfactory records that facilitated traceability of the finished quality to the brand of the paint used and hence the supplier. The organization made use of this data to choose and patronize the right supplier(s). The price and the adherence to delivery schedules by the supplier were given adequate due consideration.

(c) Rayon grade pulp manufacturing unit

The V belts are used on motors to drive the mechanism that cuts the bamboos into chips. The record of the brands of the belts used and their respective lives were maintained. The examination of the data revealed that one of the three brands which cost ten percent more had a longer life by twenty percent. The economic gain of this brand is evident. It needs to be pointed out that even if the cost of this superior brand were twenty percent more or still marginally higher, it was still worth it because of more gains accruable from ten percent less time and effort needed for the replacement of the belts. In addition corresponding loss in production is avoided. With this kind of justification, none can raise a finger on the choice of the supplier who is apparently costlier. This experience, hits the nail right on the head. It cautions against the blind policy of accepting materials on the basis of lowest quotation alone.

5.5.6 Stratification with Respect to Customer

(a) Electrical fan manufacturing unit

The term customer is used in a broader sense to include dealers and sub-dealers. A fan manufacturer, reputed for its quality was suddenly flooded with returns and complaints. The returned fans were thoroughly inspected and dismantled if necessary, to verify the veracity of the complaints and the faults reported. Some of the fans were in excellent condition and some others appeared to have been intentionally tampered with. These were not normal damages experienced in transit or handling. The returns and complaints were classified by regions and within a region dealer wise. The reason was traced to a dealer who had shown high sales to claim more incentive or bonus. Not only that, he had even let out the fans on hire, made easy money during the season and later returned the fans, as of poor quality, which on the top of it had been procured on credit. The organizations try to identify and choose right suppliers. It seems now that, conversely, the suppliers practice the same principle to deal with the 'right' organization. The policy to fool the opposite party can prove fatal in the long run.

5.5.7 Stratification by Process Stage

(a) Pharmaceutical company

A pharmaceutical company engaged in manufacture of a baby tonic suddenly encountered the problem of the presence of glass speks in the sealed and labeled bottles during final inspection. This non-conformity was too serious to be ignored. All bottles were thoroughly inspected and only bottles free from speks were marketed. The team designated to resolve the problem met and discussed the issue. The majority view was that the weak neck of the bottle was leaving speks during the stage of sealing operation for the cap. Other conjectures pointed to the possible originating sources as hardness of cap and the process of washing besides host of others. The study consisted of following a sample (batch) of bottles at all stages of the entire process - starting from their receipt in the stores through opening and cleaning, washing, filling, sealing to labeling and packing. The bottles were inspected at each stage for presence or otherwise of the glass speks. The inspection procedure used magnifying glasses and the area was adequately lit. The washing stage was identified as the factor contributing to the problem. The washing process was observed minutely. The bottles rolled over the rollers and the brushes inside the bottles, all immersed in the hot water container. The metallic end of the brushes was hurting the necks of the bottle to produce glass speks. These in turn were getting into the bottle due to centripetal force. The design of the brushes was modified. The metallic part of the brush was covered with rubber sleeve. This eliminated the problem.

5.5.8 Conclusion

The variety of above success stories substantiates the claim of the tools being simple, quick and cost effective. The application of the concept of *stratification,* among these, is simplest, quickest and costs least. This tool is versatile and universal in application.

Its application on the past data should be first step in the direction of reaching the solution to the problem. It gives a head start. It helps in identifying the likely dominant factors and their levels with favourable and adverse effects. One can choose the right combination, accept this as the standard and adapt it to reap the expected gains. The least it does is, it carries the study a step forward in search of the solution to the problem. The standards need to be revised, as more and more experience is gained. In addition to its tangible returns, the intangible rewards include the boosted moral of the human resource.

Lastly, it is documented in several books, that in every factory there is an invisible or hidden factory that keeps on producing losses. The reference is to the produce of non-conformities, that create extra non value addition work(s) resulting in huge loss of material and allied resources consumed as also loss of production potential.

What is possibly not documented hitherto, is the fact that there is a CORNER OF EXCELLENCE in every factory or work place. It implies that every factory on some day, in some shift, on some machine, some worker(s) using some material, instrument(s) and allied facilities did produce a superb piece. The phrase CORNER OF EXCELLENCE in this context means a stratum, from among the many strata possible by classifying the data by levels of the several factors listed in stanza 5.4 above, that gave rise to this exemplary performance. Now, the job of being perfect is simple, if only the data were so maintained that each output of interest can be traced to each input of concern, thoughtfully listed in the relevant cause and effect diagram. The art of stratification will help in locating this stratum or corner. The elements of this stratum should form the work standards. The entire, associated human resource needs to be oriented to follow these standards. The produce will be non-conformity free, a status superior to six sigma or ppm status being craved for ! What else do we want? Is it not an indispensable tool?

6

SCATTER DIAGRAM

6.1 DEFINITION

The dictionary meaning of the word 'scatter' is sprinkle, dissemination or dispersion.

In the present context:

For quality parameter of interest, there is need to assess the nature and or the degree of impact on the 'quality parameter' of the resulting output as reflected by the observed changes caused by the corresponding variations in the 'input parameter'. The former is called the 'dependent' variable and the latter 'independent variable'. Attempts are made to get the yield or the behaviour of the dependent variable in desired range by monitoring the causing or independent variable, in appropriately predetermined limits as indicated by the nature of the relationship between the two. The dependent variable is plotted along 'Y-axis' and the independent along 'X-axis'. Such a diagram is called scatter diagram. See Figure F6.1.

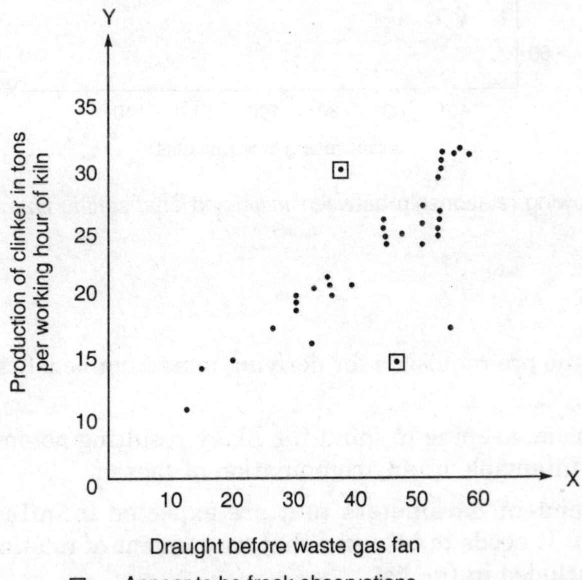

Figure F6.1 A scatter diagram showing *production of clinker in tons per working hour* versus *draught before waste gas fan* in a dry process cement plant.

6.2 UTILITY

It is *used* to indicate whether a pair of the parameters or the variables considered technologically mutually interdependent, are really so in the process under examination. If so, the nature of dependence—linear or non-linear and the degree or strength of the relationship.

Thus, it is useful in identifying the vital few parameters that play a dominant role in the final value of the product parameter of interest. It also indicates the approximate advantageous level of the process parameter to be aimed at, as also the direction for further attempts for achieving higher targets.

It also helps in substituting a cumbersome time consuming existing test procedure by an alternative less time consuming or an expensive one by an alternative cheaper method, yet equally reliable one for speedier information necessary for feedback for effective control of process and or product parameter. An example of this appears in Figure F6.2

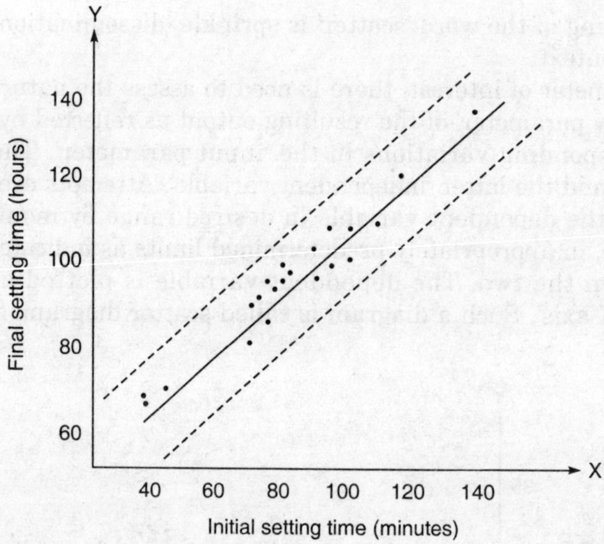

Figure F6.2 Showing relationship between *initial* and *final setting times* of cement.

6.3 PRE-REQUISITES

In industrial situations, the pre-requisites for deriving maximum benefits from the use of this technique are:

Selection of the problem, keeping in mind the likely resulting potential benefits—direct, indirect, tangible or intangible or any combination of these.

Listing of all independent parameters that are expected to influence the dependent parameter of interest. It needs to be noted that assessment of relationship is not possible for the factors not included in the list.

Care should be taken that such factors are by and large measurable or at least quantifiable and controllable. The unit of measurement should preferably be less than one-tenth of the observed range or the specified tolerance whichever is lower. Like wise the repeatability and reproducibility errors together should not exceed one sixth of the observed range or the specified tolerance.

6.4 PROCEDURE

The approach consists of collecting a set of 25 observations on corresponding dependent and independent variables or parameters of factors of concern. Care needs to be taken that the data are spread over a period of time that gives a fair chance of occurrence or representation to the factors that are beyond any feasibility of control. The data are then plotted on a graph to draw scatter diagrams, of dependent variable versus each of the independent variable for interpretation. The scale for plotting the graph or the scatter diagram may be so chosen that the spread covers an area close to a square and large enough to be adequately visible.

A few vital, among these, that indicate relationship worthy of exploitation, are then chosen for determining the ranges between which these should be controlled to get the dependent parameter or variable in the acceptable range. The control is exercised by monitoring through samples of suitable size at appropriate intervals. If necessary, these are taken up for further in depth study to assess multiple correlation and the likely *impact* with the help of more advanced techniques.

6.5 COMMONLY OBSERVED SCATTER DIAGRAMS

The Figures F6.3 (a) to (d) are examples of commonly observed scatter diagrams that *fail to provide evidence of any worthwhile relationship*. This is so, whenever, the points are so scattered that they are spread over a square, circle, ellipse or rectangle (vertical or horizontal).

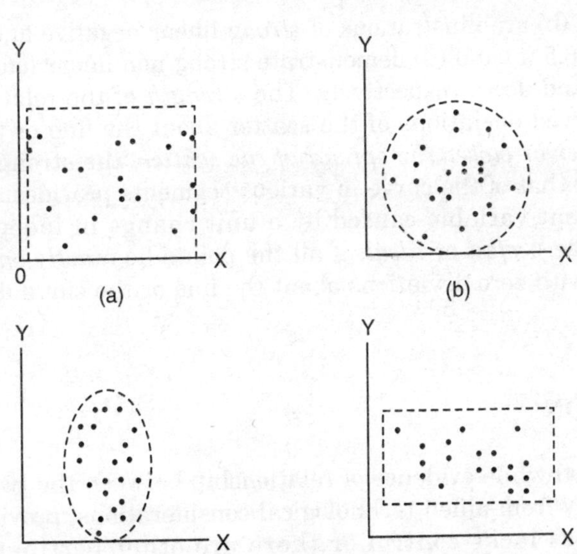

Figure F6.3 Showing scatter diagrams indicating *no worthwhile relationship*.

The Figures F6.4 (a) and (b) provide evidence of *weak linear positive* relationship. On the contrary the Figures F6.4(c) and (d) indicate the presence of *weak linear negative* relationship. The relationship is termed *linear if a line can be drawn through among these to fairly represent the scatter,* as shown in these figures. It is said to be *positive if 'y' is seen to increase as 'x' increases* & vice versa and *negative if 'y' decreases as 'x' increases* and vice versa.

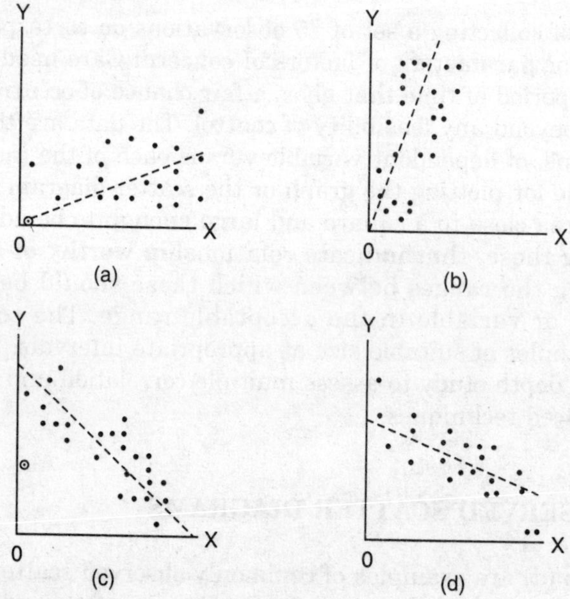

Figure F6.4 Showing scatter diagrams indicating *weak linear relationships.*

Figures 6.5 (a) and (b) are illustrations of *strong* linear negative and positive relationships respectively. Figures F6.5 (c) and (d) demonstrate strong non-linear (curvilinear) relationships with peaks facing up and down respectively. The *strength* of the relationship is judged from the extent of the observed deviations of the scatter about the line or the curve representing the same. The *narrower or closer the spread of the scatter,* the stronger is the relationship. The slope of the line or that of the curve in various segments provides a measure of the likely change in the dependent variable caused by a unit change in independent variable. The relationship is said to be *perfect* or *ideal,* if all the points lie *exactly on a straight line or any specific curve,* that is, with zero deviations about the line or the curve drawn, as the case may be.

6.6 INTERPRETATION

The absence of any worthwhile evidence of relationship between the two parameters, when it is considered very likely from allied technological considerations, provides an important clue that either the process lacks control or there are other pertinent factors which are camouflaging this. These need to be investigated and taken care of. Once this is done, the relationship which was hidden or dormant hitherto, might surface up to yield the extra information being sought. This ought to be confirmed by collecting fresh data, that is by

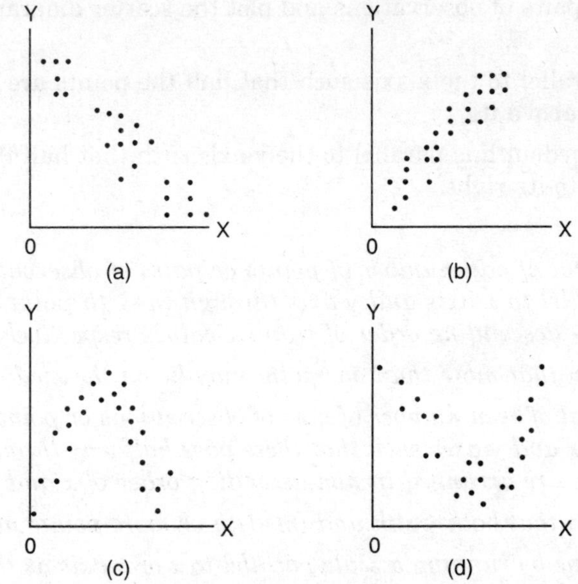

Figure 6.5 Showing scatter diagrams indicating *strong linear* and *curvilinear relationships.*

repeating the study by taking the extra indicated precautions. This assumes that the data are genuine, reliable and have desired accuracy. False data can mislead with 'disastrous' consequences.

Like wise an evidence of relationship should be viewed carefully, particularly when none is expected. These are termed *spurious or non-sense* relationships or correlations. Such situations have been known to have led to abuse of this simple yet useful tool.

Apparent anomalies should be taken with a pinch of salt. These, however, serve a very important purpose, by revealing the need to reconcile the diverging aspects—the views or opinions based on associated scientific or technical knowledge and the facts observed from the live data. The attempts to reconcile this gap between the theory and the ground reality have often led to identification of contributory factors, hitherto unsuspected. The follow up leads to the solution to the problem on hand.

A word of caution, the relationships are considered valid only for the range or the interval of the variation observed. Interpolations or forecasts within the range experienced are valid. Any extrapolation is not permissible. If the technical considerations or the analysis of data or the need justifies, the fresh data should be collected in the extended range of interest and analysed to confirm its validity before implementing the recommendations resulting from the analysis of data. It is a healthy practice to make confirmatory trials before implementing the recommendations of the findings on routine basis. *Interpret* the scatters *intelligently*.

6.7 DETECTION OF PRESENCE OR OTHERWISE OF RELATIONSHIP BETWEEN A PAIR OF VARIABLES, FACTORS OR PARAMETERS

(a) Linear

The procedure described below is followed (See Figure F6.6)

Consider about 25 pairs of observations and plot the scatter diagram. There are 26 points in the said figure.

Draw a median parallel to the x-axis such that half the points are below the median line and the other half above it.

Like wise draw a median line parallel to the y-axis such that half the points are to its left and the other half to its right.

Notes:

1. *If there are a total of odd number of points or pairs of observations, say (2n+1), then draw lines parallel to x-axis and y-axis through (n+1)th point, after these have been arranged in non- descending order of y and x values respectively.*

   It may so happen that more than one point may lie on the median line.

2. *If there are a total of even number of pair of observations or points, say (2m), then draw lines parallel to x and y axes such that these pass half way through (m)th and (m+1)th points after these are arranged in non-ascending order of x and y values respectively.*

   It may so happen that both (m)th and (m+1)th or more points may lie on the median.

3. *This is easily done by running a scale parallel to x or y axis as the case may be, till the criteria stated above are satisfied.*

   Designate the four quadrants thus formed as I, II, III and IV. Count the number of points in these quadrants respectively, say n_1, n_2, n_3 and n_4. Ignore the points falling on the median lines, if any. In Figure F6.6 $n_1 = 1$, $n_2 = 12$, $n_3 = 0$, $n_4 = 11$. Two points lie on the median parallel to y-axis

Note:

In the absence of any relationship, hypothesis of assumption of $n_1 = n_2 = n_3 = n_4$ should hold valid, except for random or chance deviations of no consequence. In particular, ideally $(n_1 + n_3)$ should be equal to $(n_2 + n_4)$. Statistically speaking, there should be no evidence of these differing significantly.

In the event of presence of any worthwhile linear relationship between the two, the difference between $(n_1 + n_3)$ and $(n_2 + n_4)$ should be large enough, wider than what can be given a benefit of doubt on account of random or chance fluctuations. For significance, under binomial assumption either should exceed $(n/2 + \sqrt{n})$ at about 95 percent level of confidence where $n = n_1 + n_2 + n_3 + n_4$.

If $(n_1 + n_3)$ is significantly greater than $(n_2 + n_4)$, then the relationship is deemed positive. On the contrary if $(n_2 + n_4)$ is greater than $(n_1 + n_3)$, then the relationship is deemed negative.

In this case, see Figure F6.6,

$(n_1 + n_3) = 1$ and $(n_2 + n_4) = 23$;

$n/2 + \sqrt{n} = 13 + 4.9 = 17.9$, $23 > 17.9$

Thus there is very strong evidence that the 'Tensile strength' is linearly related to 'percent carbon equivalent', and the former can be achieved through control of the latter in appropriate range provided other factors are not disturbed and other conditions remain the same. Also, this relationship can be used to predict Tensile Strength given the value of Percent Carbon Equivalent.

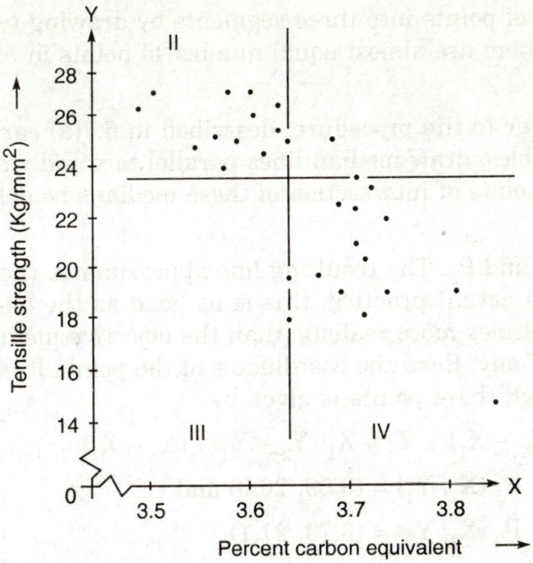

Figure F6.6 *Detection of linear relationship* between two variables, factors or parameters.

(b) Non-Linear

To verify presence or otherwise of 'specific' non-linear relationship, standard transformations, available from published literature are made such that the transformed data exhibit linear relationship, which in turn can be verified with the help of procedure described above. Standard statistical computer soft wares can also be used to obtain solutions to large or more complex problems.

6.8 EVALUATION OF LINEAR RELATIONSHIP AND COMPATIBLE SPECIFICATIONS

(a) Determination of the equation of the line

Objective Method: Let there be 'n' pairs of observations for 'x' and 'y'. Then, the regression line is given by y = a + bx, where 'b' is the regression coefficient and 'a' the constant, called the intercept on y-axis.

The constants 'a' and 'b' are so determined that the sum of square of deviations between the observed values and the corresponding estimated values, from the above equation or read from the relationship graph, is least. This is why, this line is called the line of 'best fit'. The differentiation procedure for minimizing, estimates the values of the constants 'a' and 'b' as under:

$$b = \left[\Sigma(x-\bar{x})(y-\bar{y})\right]/\Sigma(x-\bar{x})^2 \text{ and } a = \bar{y} - b\bar{x}$$

Alternative, simple procedure is given below: (see Figure F6.7)

Divide the scatter of points into three segments by drawing two vertical lines parallel to y-axis, such that there are almost equal number of points in each of these segments A, B and C.

On the lines similar to the procedure, described in 6.7(a) earlier, to detect relationship between two variables, draw median lines parallel to x and y axes for points in segments A and C. Let the points of intersection of these medians be designated P_1 (X_1, Y_1) and P_2 (X_2, Y_2).

Join the points P_1 and P_2. The resulting line approximates the line of 'best fit', called the regression line. In actual practice, this is as good as the one derived by the objective method. It is sometimes more realistic than the objective method because it takes care of the freak values, if any. Read the coordinates of the points P_1 and P_2. The equation of the line passing through these points is given by:

$y = x (Y_2 - Y_1) / (X_2 - X_1) + Y_1 - X_1 (Y_2 - Y_1) / (X_2 - X_1)$

In Figure F6.7 $\quad P_1 (X_1, Y_1) = (3.59, 26.0)$ and

$P_2 (X_2, Y_2) = (3.73, 21.1)$

Thus the equation of the line is $y = -35x + 151.65$

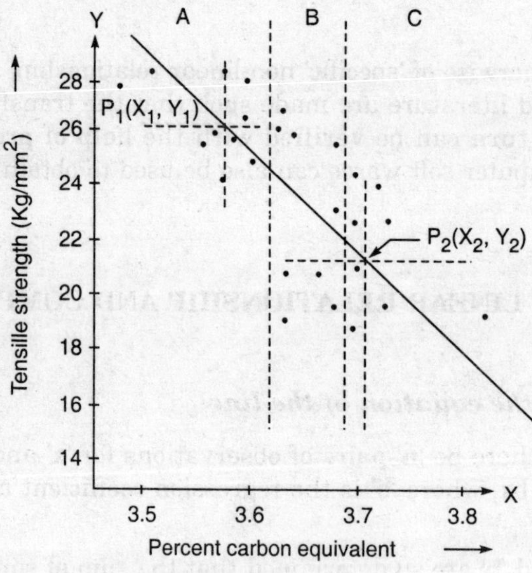

Figure F6.7 Determination of equation of the regression line.

(b) Estimation of the error

The estimate of the residual error (S_0) is given by : $S_0 = \sqrt{[S (y - y^*)2 / (n-1)]}$

Where y and y* are the observed and estimated values of y respectively for given x.

The estimate of the residual error is useful in determining the specification limits of the independent variable x for given specification limits of dependent variable y. To find these:

draw lines parallel to the line of best fit at twice the distance of the estimated standard deviation, for 95 percent confidence.

Given upper (U) and lower (L) specifications, of y, the corresponding specifications for x, that need to be met may be determined as shown in Figure F6.8. Of course if higher y is preferable through further increase of x, keep on trying it, till it reaches the convex peak. The advantage is found in the relaxed control of x to get simultaneously y even in narrower limits. See the next stanza.

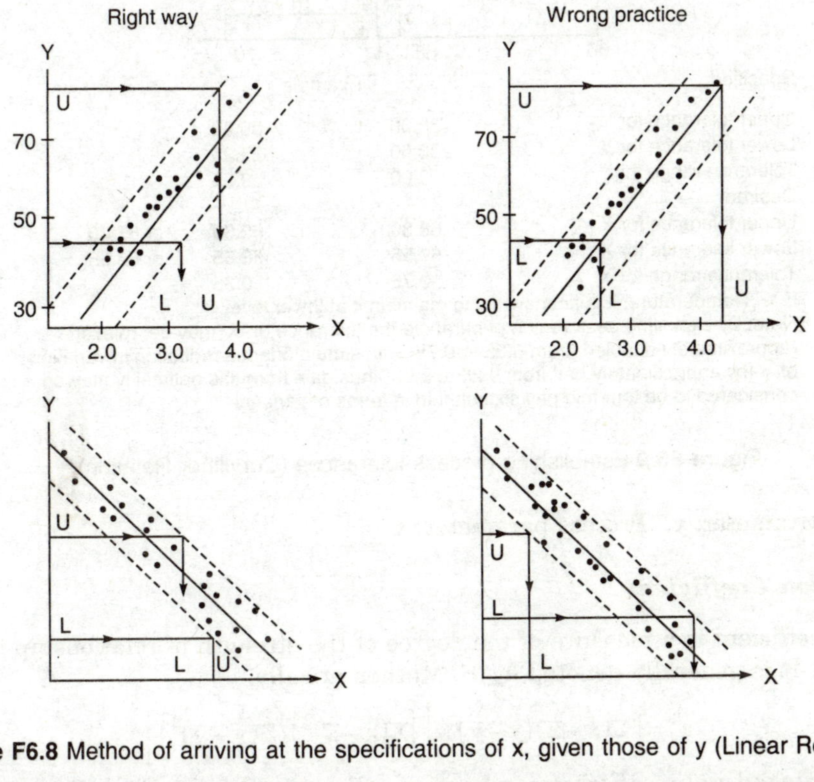

Figure F6.8 Method of arriving at the specifications of x, given those of y (Linear Relation)

However, if the relationship is non-linear, say parabolic (see Figure F6.9), the specification of y should be chosen close to its peak, that is, the maximum value of y should correspond to the point of intersection of the tangent with the y axis. For a given specified range of y, the range required for control of x, the corresponding process parameter, shall be the widest possible under the circumstances and therefore least rigid control for greater economy.

The criteria may be carefully noted for the positive and negative relationships. The wrong practices, *generally in vogue out of ignorance,* must be understood and avoided. It may also be understood that feasible limits for x are not possible if the vertical distance between the dotted parallel lines indicating natural bounds of variation around the regression line or path or curve is greater than the specified tolerance—the interval between the specified upper and lower limits, that is if the Residual Variation of y is greater than the specified Tolerance Range of y. It must also be appreciated that the line or curve predicting x given y is not the same as that for predicting y given x. The deviations for prediction d_1 and d_2 respectively are not the same.

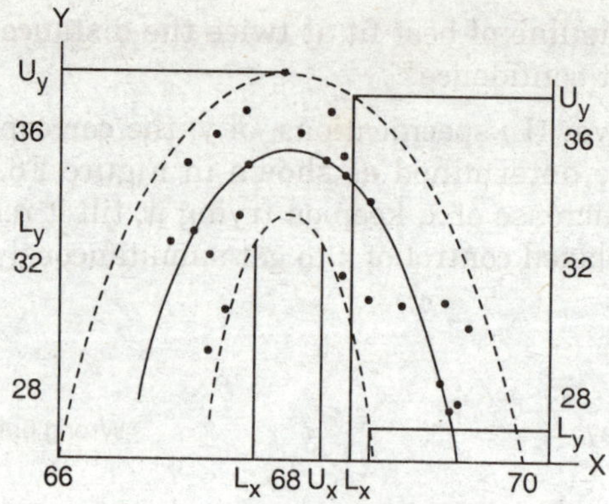

Specified	Example 1	Example 2	
Upper tolerance for Y	38.00	36.50	
Lower tolerance for Y	33.00	27.25	
Tolerance range for Y	5.00	9.25	
Desired			
Upper tolerance for X	68.30	69.00	or 67.00
Lower tolerance for X	67.55	68.65	or 66.65
Tolerance range for X	0.75	0.35	

If X is temperature, it is economical to maintain it at lower level.
Note: By exploiting peak region of parabola the tolerance of 'X' may be, relaxed (approximately doubled, from 0.35 to 0.75) with simulataneous reduction in variation of Y (by approximately half from 9.25 to 5.00) thus gain from this optimality may be considered to be four fold nay sixteen fold in terms of variance.

Figure F6.9 Establishing process tolerances (Curviliner Relation).

Product parameter: y, Process parameter: x

(c) Correlation Coefficient

Correlation coefficient is a measure of the degree of the strength of relationship between the two variables. It is generally denoted by 'r'. Mathematically,

$$r^2 = \left[\Sigma(x-\bar{x})(y-\bar{y})\right] / \left[\left\{\Sigma(x-\bar{x})^2\right\}\left\{\Sigma(y-\bar{y})^2\right\}\right]$$

The calculated value of 'r' can sometimes be very misleading. For example, see Figures F6.10 (a, b, c, and d), wherein r = 0.83. Figure (a) reflects a situation commonly experienced in industry when two variables are fairly well related amenable to the procedures and interpretation discussed herein. Figure (b), on the contrary, clearly indicates that the relationship is not linear and the optimal value is indicated by its peak. The linear relation in Figure (c) will undergo a major change, if the freak point shown in square is ignored. The slope of the line will reduce and all the rest of the points are seen to lie perfectly on the straight line. Thus, the corresponding change in y for unit change in x will be less. Simultaneously the value of 'r' will change from 0.83 to 1.0, indicating a stronger relationship, strongest or perfect in this particular hypothetical example. In contrast, the linear relation in Figure (d) will vanish, if the freak point shown in the figure is ignored, implying that even perfect control of x is unlikely to be of any help whatsoever in reducing the variation of y in this particular hypothetical example, indicating the need to search for other factors, that might matter. It is not unlikely that the factors, least suspected on the basis of available technical knowledge, have been innocently ignored to such a large extent that (the process variation of this parameter has increased substantially over what it used to be earlier) these have become important.

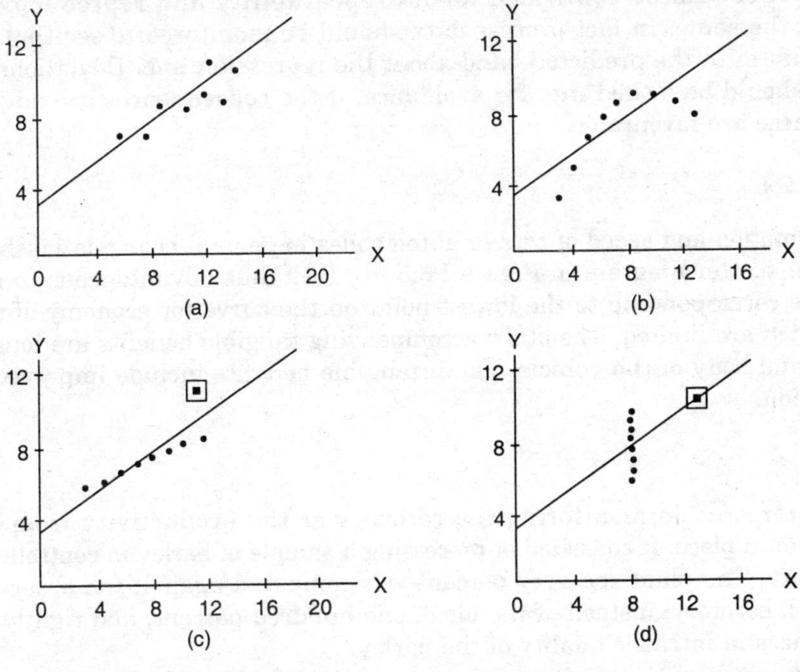

Figure F6.10 Hypothetical scatter diagrams that caution from pitfalls.

Such pitfalls can be considerably avoided by plotting a scatter diagram and looking at the pattern of the spread of the points with a special eye on freak points. In fact the procedure given for evaluation of linear relationship and compatible specifications for drawing a regression line, has a built in system, to ignore the misleading contribution from freak points. The pair of values corresponding to freak points, observed if any, should be ignored and 'r' calculated afresh, from the rest of the homogenized data only. Thus, the scatter diagram is an indispensable tool, for valid inferences, whenever the relationship between a pair of parameters merits consideration.

6.9 SUCCESS STORIES

(a) Jute Mill

The motors are used to run the looms. As the speed of the drive motors is increased, so is the net speed of the looms or production increased. However, beyond certain stage there is slippage of belts and the net speed is adversely affected. A scatter diagram of net production versus speed of motors was plotted. It resembled the shape shown in Figure F6.5(c). The optimal level corresponding to the peak prescribes the most advantageous speed at which the motors ought to be run.

Like wise, the optimal levels with respect to speed and breakage rate in spinning was examined. It needs to be noted that these optimal levels are very likely to differ from one factory to another, depending on many factors such as raw material, environments and maintenance of machines. Therefore, utmost care should be taken to assess the nature of relationship. It should be confirmed for its repeatability and reproducibility, before implementing the same. In fact process data should be monitored to see that the current operations conform to the predicted band about the regression line. Deviations beyond the band, if any, should be looked into for avoidance, if the consequences are adverse and for retention, if these are favourable.

(b) Automobiles

The fuel consumption and speed of cars or automobiles in general bear relationship similar to those shown in scatter diagram in Figure F6.5 (d). It is thus advantageous to run these at optimal speeds corresponding to the lowest point on the curve for economy of the fuel, the reserves of which are limited. The other accompanying tangible benefits are longevity of the tyres, engine and body of the vehicle. The intangible benefits include improved safety and reduced pollution.

(c) Brewery

A brewery, interested in monitoring the recovery or the productivity from barley, had elaborate system in place. It consisted of processing a sample of barley in controlled conditions in the laboratory. The plant recovery percent was compared against the observed recovery percent in the laboratory, instead of the ideal, one hundred percent, and rightly so, to allow for the differences in intrinsic quality of the barley.

A scatter diagram of the plant recovery percent versus the laboratory recovery percent was plotted. It resembled the Figure F6.3(c). It was very revealing and surprising to the management to know that:

the two recoveries did not bear any relation between themselves, and

the process variation was more than that observed in the laboratory, no second thought, it is just a known expected phenomena.

This lack of relationship, does not reveal (rather conceals) the factors for high variation in the plant process recovery. Thus the objective for which elaborate laboratory was set up at huge cost and the introduction of test system appeared to have been defeated.

It pointed out the need for looking into the reasons for this among other associated process factors, which are camouflaging the expected relationship, including accuracy of data—the method of estimating the plant recovery and testing in the laboratory encompassing the use of chemicals (their concentration) and procedure (such as durations). May be, the standard practices have been laid down, but are not being adhered to. If not, these need to be standardized.

(d) Bicycle

It was observed during the assembly operations of the bicycle that the pedal was hitting the chain stay instead of having a free movement, even though only the accepted components were being assembled. The *inspection department* blamed the *design department* for its faulty design. The design department in turn blamed the inspection department, which had taken the entire responsibility of assuring that the process conforms to the laid down procedures and the product satisfies the specified dimensional tolerances. The data on the magnitude of interference and the dimensions of the suspected parameters of the components involved were obtained and scatter diagram plotted. It resembled the Figure F6.3(a). The data were small, yet conclusive. The interpretation was explained to the design personnel; that even if all the components were perfectly alike, the problem is likely to show up. They agreed to look, for the problem, elsewhere.

The case examples discussed in sections (c) and (d) above affirm the utility of analysis of factual data (scatter diagram is, one such analysis), to think rationally to resolve the problem rather than engage in the game of blaming each other. The game of blaming vitiates the entire environment, detrimental even for the survival of the organization.

(e) Medical Research

Curative power of medicines for high blood pressure (B.P.) is being judged by relating the improvement in B.P. over time after administering the prescribed dose. However, such studies when conducted along with control group, consisting of patients who have not been treated, and compared with the former did not provide good enough evidence of the efficacy of the treatment. Thus there is need to conduct such valid statistical studies to avoid any hasty raw conclusions.

(f) Steam Boiler

Steam boilers are used to generate steam required for various processes in textile, vanaspati, pharmaceuticals, laundry attached to hospitals or hotels, rayon grade pulp, thermal power houses, starch and the like.

Common worthwhile pertinent relationships, that appear to be relevant for regular monitoring, for control and improvement of productivity of the boiler house and the steam are:

The amount of the steam generated versus fuel consumed and

The amount of finished product or suitable weighted index of the product range versus the amount of steam consumed

The above scatter diagrams were drawn for a pharmaceutical company. Once again the likely relationship was not apparent. Generally, such anomalies are attributed to erratic variation in the quality or ash content of the coal or the fuel received. The discussion among the personnel of the production and boiler house led to the following standard practices for compliance:

Production Department will communicate their requirement of steam or lack of it to the boiler house a couple of hours in advance, to enable them to plan steam generation to avoid wastage of resources and

*Boiler house will arrange to feed coal in appropriate small broken sizes instead of huge lumps and at appropriate intervals to ensure proper combustion that in turn shall help maintain desired pressure and supply.

These measures reduced coal consumption for the same production by 35 percent. Financially, a saving of only one percent in a plant of normal size of sugar, textile or vanaspati will have the same implications.

Next, the relationship between the coal consumed and water consumed (used as an indirect measure of steam generated in the absence of suitable steam meter that will take care of varying temperature and pressure) was established. Such a relationship could not be established in the new plant. Investigations revealed a fault in the design. The capacity of generation was not compatible with the needs, leading to process failures causing production jams. The design was appropriately modified.

(g) Shock absorbers of railway goods wagon

The shock absorbers consist of metallic discs embedded with circular rubber strips on either side (there are holes in the discs for proper grip of rubber strips on either side) placed in a cylinder. These can be seen at the ends of the wagon, at platform level, parked at the railway stations. There was substantial amount of rework for too porous or too hard rubber caused by less or excess rubber fed during assembly of discs before curing. The control practice in vogue was to cut strips as per standard length depending on the designed parameter of circles or rings on the discs. This apparently good practice was found to be unsatisfactory because the net amount of rubber going into the process was not uniform, since rubber shrinks over time and the interval between drawing of the strip and subsequent processes up to curing was not uniform. A relationship between length of the strip and its weight was established at the drawing process stage. This relation was used to convert standards on length to weight. This exercise, would have been redundant, if the process sequence permitted cutting of the strip into pieces of desired length at this stage and stored as such till next stage of assembly with discs. For ease of operation, go/no go limits were fixed for weight for various requirements based on the corresponding linear relation which resembled the shape shown in Figure F6.5(b). Both, the quality and productivity improved simultaneously.

(h) Vanaspati

One of the final parameters of hydrogenated oil monitored by the manufacturer is its *final colour.* It is likely to be influenced by the initial colour of the oil and the quantity of bleaching agent used as also its brand or quality. Reprocessing of batches to correct for non-conformity of colour or their diversion to different market segments was a routine phenomena. The ratio of quantities of bleaching agent used and oil was standard and constant over batches. Past data were used to draw scatter diagrams with final colour on y axis and initial colour on x axis, separately for imported and indigenous bleaching agents also called activated carbons. Both these indicated fairly good linear relationships. The line for indigenous bleaching agent had greater slope, indicating that less quantity of indigenous agent was required against the imported. However, the residual error for the indigenous variety was larger in relation to the imported, indicating the need for more homogeneous production of the bleaching agent by the local supplier. Had one common diagram been drawn for both the varieties, the possibility of the relationship being camouflaged or concealed is not ruled out. This, incidentally, is an example of supplier wise stratification. Therefore, wherever appropriate and possible, stratified or group wise scatter diagrams should be drawn.

(i) Heavy electrical

A heavy electrical equipment plant manufacturer was facing the problem of maintaining dimensions of *rotor slot trough,* as per drawing. A list of the relevant influencing factors was prepared. These included process parameters and the dimensions of glass cloth and die. The scatter diagrams of output versus input parameters were drawn one by one. This helped in identifying the dominant parameters needing control.

This attempt also indicated when the die became due for recalibration for its dimension(s). Hitherto, in the absence of any knowledge, the die was unnecessarily tempered with, though in good faith, causing loss of effort, material and production.

(j) Chemical plant

A plant was manufacturing Titanium Dioxide. The final stage of process consisted of feeding slurry at the top of a rotary kiln and getting the finished product at the bottom. The parameters of interest were quality and productivity of the output ; moisture and feed rate of the slurry; and temperature of the kiln in three zones.

Six scatter diagrams of Productivity per unit of time versus Quality; Moisture of Slurry; Feed Rate of Slurry; and Temperatures T_1, T_2 and T_3 in the three zones of the kiln were plotted. Adequate past data were used.

It was a pleasant surprise to learn that productivity and quality bore positive relation. Thus it was sufficient to study the behaviour of five input process parameters. Moisture and feed rate did not seem to make any difference in the prevailing range of operational variations, implying that these were conforming to valid standards. There was strong relationship with all the temperatures. The scatter diagrams among the temperatures, plotted subsequently, taken two at a time, also showed positive relationships among themselves. It was decided to monitor temperature in the middle zone of the kiln only. The scatter indicated reduction of the temperature by about 60 degrees for good results. The management including technocrats were hesitant to accept the suggestion for fear of adverse consequences. The possibility of even an explosion in this delicate chemical process was not ruled out. They were persuaded with the supporting evidence from the data that fears were not justified. A compromise was, therefore, struck. It was agreed to exercise control of the temperature in narrower limits, through extra vigilance and simultaneously reduce temperature by 10 degrees at a time and reach the target in stages, if no hurdles are encountered. This approach is synonym with evolutionary operations. The implementation of this strategy resulted in record production of better quality of Titanium Dioxide accompanied by reduced fuel consumption. The tremendous gain in productivity of indirect costs can be imagined. That year the production bonus too was highest ever.

The approach has been profitably applied in cement industry too, where the kiln operations are similar.

(k) Rubber Manufacturers

A rubber manufacturer was rolling a *tread* to be shaped and cured into a cycle tyre. The cross section of the tread, along its width, looked as shown in Figure F6.11. Its weight per meter was an important parameter for control. The procedure consisted of cutting one meter length at regular intervals of time, checking its weight for conformance and making necessary corrections in the process, if warranted.

The width, thickness, length and density of the rubber compound together determine the weight. The compound and its density were found to be fairly standard. The control of the width too was adequate. Samples of standard lengths were cut. Thus only thickness was desired to be investigated.

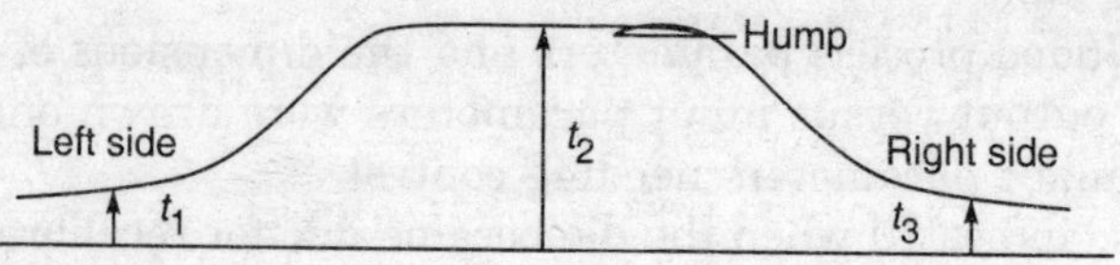

Figure F6.11 Showing cross section of tread of cycle tyre.

The thickness at three positions, one each at the ends and one in the middle along the width of the tread were associated parameters of concern. The process data were collected on the samples being tested for weight (w) and on thickness at the three positions (t_1, t_2 and t_3). Six Scatter diagrams were plotted—w versus t_1, t_2 and t_3; t_1 versus t_2 and t_3; and t_2 versus t_3. It was seen that t_1, t_2 and t_3 were related among themselves. It was therefore possible to forecast weight (w) and exercise its control by monitoring thickness at any one of the three positions. The repeatability and reproducibility errors of measurement of thickness of rubber sheet is more than that for solid sheets. Also, the more the thickness of the rubber sheet, the more is the error of measurement, inspite of the special flatted ends of the micrometers designed for use in rubber industry. Thus the revised system for control of the process consisted of a thorough first off inspection to ensure parallelism of the rollers and the gap between the two and subsequent periodical check for thickness only at position t_1. Occasionally position t_2 was checked to re-affirm the parallelism. This procedure avoided periodic cutting of the tread and thereby the end waste during cutting of the tread into pieces of length equal to the periphery of the tyre prior to its insertion into moulds for curing. The bonus was, the reduced handling.

(l) Conclusion

The foregoing case examples are not intended to contradict the technical knowledge available. These are intended to support; by highlighting each case on its merit; that there are some factors, over which one has no control and there are others, with which one can play. It is necessary to assess the impact of changes in the latter that will counter the adverse effects of the former to reap the accompanying benefits. As discussed earlier, the expected relations may get dormant and invisible because of excessive variations. In such situations too, the support is provided by bringing forth need to exercise better control by reducing variations in the factors. Scatter diagram is a tool to aid the technocrat to perform better.

7

HISTOGRAM

7.1 DEFINITION

French statistician A M Guerry is credited with development of *histogram* in 1833.

A Histogram may be defined as graphical cum pictorial presentation of variation of a single parameter observed in the data emerging from any process. Generally the horizontal x-axis represents the measurable value at appropriate scale and the vertical y-axis the frequency with which the number of times the value has occurred in the form of a vertical rectangle as illustrated in the Figure F7.1.

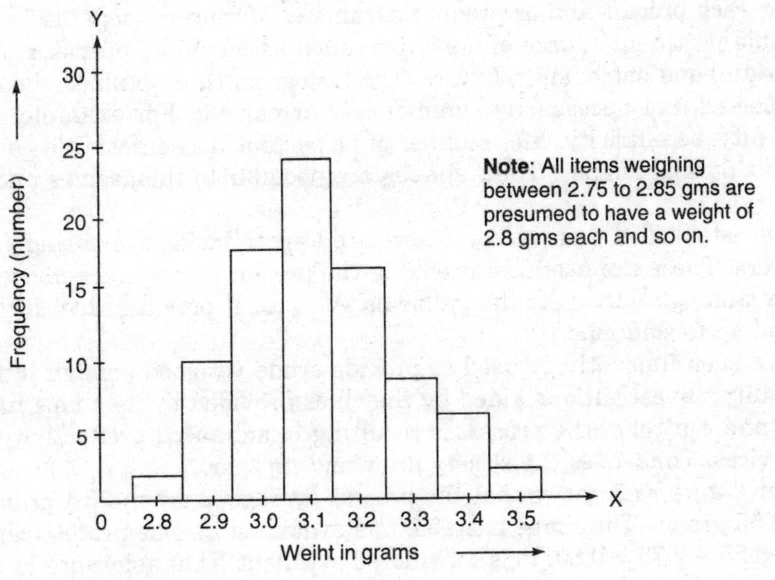

Figure F7.1 Histogram of weight of tablets in grams.

7.2 THE ART OF MAKING IT

The measurements should be made preferably with least count approximating to one tenth of the specified tolerance or range of the observed spread over stable period of the process. In such cases, as mentioned above and as shown in Figure F7.1, the value is shown on x-axis and the frequency along y-axis in appropriate scales. This should do for most of the industrial situations.

It is of course pertinent, that the data should be relevant, current, true, authentic and adequate.

7.3 UTILITY

It is good to remember that all data are subject to variation. It is necessary to understand the factors that cause, the observed variations. The sources of variation are peculiar to each process and hence the product. Different causes give rise to different patterns. It is helpful to understand their effect. We are then in a better position to conduct systematic investigations; to identify and trace the root causes of unwanted and avoidable variations ; and to search for their remedial measures. In addition to providing help in finding solutions to the problems, it cuts down on the time required to do so. The proposed remedies should be confirmed, suitable system evolved and implemented to sustain the gain with the help of run chart, the subject matter of the next chapter. Adequate education and training needs to be imparted to the workforce to avoid hiccups. Simultaneous study of short term and long term variations is generally more informative, eye opener, educative and helpful in reducing and controlling the same.

To take quantum advantage from studies speedily, it is extremely important to make histograms for each process and or product parameter of concern, separately for each known and controllable pertinent source of variation, such as machine, operator, material source (supplier or origin) and batch, shift, fixture. The histogram thus obtained should be compared with that expected, not necessarily symmetrical or normal. For example; run out, taper, ovality, centrality, eccentricity, life, number of or percent nonconformities in small samples will not exhibit normal shapes. Their shapes are peculiar to themselves and bear befitting names.

These have established their utility. These are easy to make, convincingly communicative even to doubters. These are useful in breaking the proverbial ice and remove the hurdles in the progress of finding solutions to the problems. At least, it provides clues for further probing ultimately leading to solutions.

These have been innovatively used to provide crude yet good enough initial estimate of process capability. Investigations aided by the clues provided by its shape have often led to improvement and control of the processes resulting in enhanced productivity and quality of goods and services. Thus, does the society grow and prosper.

A glance at Figure F7.1 shows that the process average is around 3.1 grams and it varies from 2.75 to 3.55 grams. The shape is stable and symmetrical. The process capability may be estimated as: $3.55 - 2.75 = 0.80$. It is considered excellent if the tolerance is 1.00, more so if the specifications are 3.10 ± 0.50; and poor if the tolerance is narrower than 0.80. On the contrary, if the specified mean is different from 3.10, the situation can be easily remedied by appropriate action to get the desired mean. In this case, the feed rate of the mix may need adjustment.

These procedures, often over estimate the spread or under estimate the capability of the process. Appropriate methods to collect data supplemented by homogenization, that helps to alienate the contributions from assignable causes if any provide more realistic estimate of the process capability. We also, generally, under estimate our capabilities and potential. Thus, we miss many wonderful platinum or diamond, life time opportunities.

7.4 PRE-REQUISITES

Before we proceed to discuss its applications, it would do well to recall that the quality has been universally defined as *fitness for use*. The fitness needs to be translated into *product specifications* encompassing physical, functional and reliability requirements. These in turn, require evaluation of compatible *specifications of* parameters of various inputs including *process(es)*. The *specifications* comprise *optimal target* and acceptable pattern of spread or *variation around it.* The lesser the spread, the better it is.

As a first step, in any given situation, it is important to understand and list the factors or causes of variation and divide these in to two distinct groups. One, over which one has no control ; usually termed chance, common, natural or random causes. The other, which we can manoeuvre, manipulate or control at will; usually termed assignable, special, unnatural, or systematic causes. The amount of variation, observed when the process is under the influence of former set of causes alone, is considered inevitable and is called process capability. It is estimated as six times sample standard deviation. Any increase in this is attributed to the latter and is economically avoidable.

As a next step, the pattern of variation when the process is under the influence of chance causes alone called state of statistical control, is familiarized. These are then compared with those obtained from the live process data. Remedial measures are contemplated and executed, if the differences between the two are considered too wide to be palatable. Additionally, as a positive measure, studies are planned to improve quality through continuous efforts on reducing the variability.

Experience has shown that, often about 50 observations are considered adequate for right inferences. However, one should be flexible depending on the situation. The considerations take into account, desired accuracy, economics and feasibility. The latter in turn depends on the rates of production and inspection, their costs, equipments and the skills required. More than 100 observations are mostly redundant.

7.5 SUCCESS STORIES

(a) Machine tools

A machine tools, engaged in job shop production, found conventional run or control chart impracticable, since the production was over by the time meaningful feed back on conformity or otherwise of the dimensional parameters is received from the *standards room* responsible for measurement. A job shop is defined as a job of a few pieces at a time repeated rarely or a job consisting of a large number of pieces but accomplishable in a short duration, say at most a couple of hours. The job being referred here belongs to the latter category. Innovative application consisted of setting up the machine and allied process parameters at satisfactory levels and running the machine to produce just about 50 pieces. Histograms of all dimensions of interest are made. If these looked like the one shown in Figure F7.1 and properly centered around mid value of the specified limits, the machine was run to complete the planned production. The last about 50 pieces are collected separately and like wise checked for conformance with the help of histograms. If the pattern of the histograms both at the commencement and completion of the job are acceptable, then the entire lot is accepted and transferred to assembly. This assures right quality and simultaneously avoids one hundred percent inspection. It is known, from the previous experience, that the machine runs stably for the duration of the job. Hourly or half hourly outputs, as deemed appropriate, may be collected in separate trays or bins to minimize the additional inspection work load that might become necessary to segregate conforming and non-conforming pieces, in the event of any unlikely failure of the system during the scheduled run. This is an *innovative use* of the *histogram* for *control of the process*. Its *usual role* is that of a *post mortem to verify conformance* to desired form. Even so, it provides invaluable guidance for immediate future use for improvement and control.

(b) Glass shells

Let us now look at the Figure F7.2. This summarizes the production per shift of 8 hours duration, by a team of a dozen workers engaged in a particular synchronized sequence to produce glass shells. The process is labour intensive. It is amazing that inspite of apparently identical conditions the production has varied from a low of 340 to a high of 580 (about 80% more than the low value of 340), with an average of about 433. There are three peaks, one each at 370,410 and 490 besides a freak (yet feasible and repeatable performance level) pair of bars at 550 and 570 with average of 559. Imagine the gains of production, in the light of the fact that the raw material cost, which is directly proportional to the number produced, is

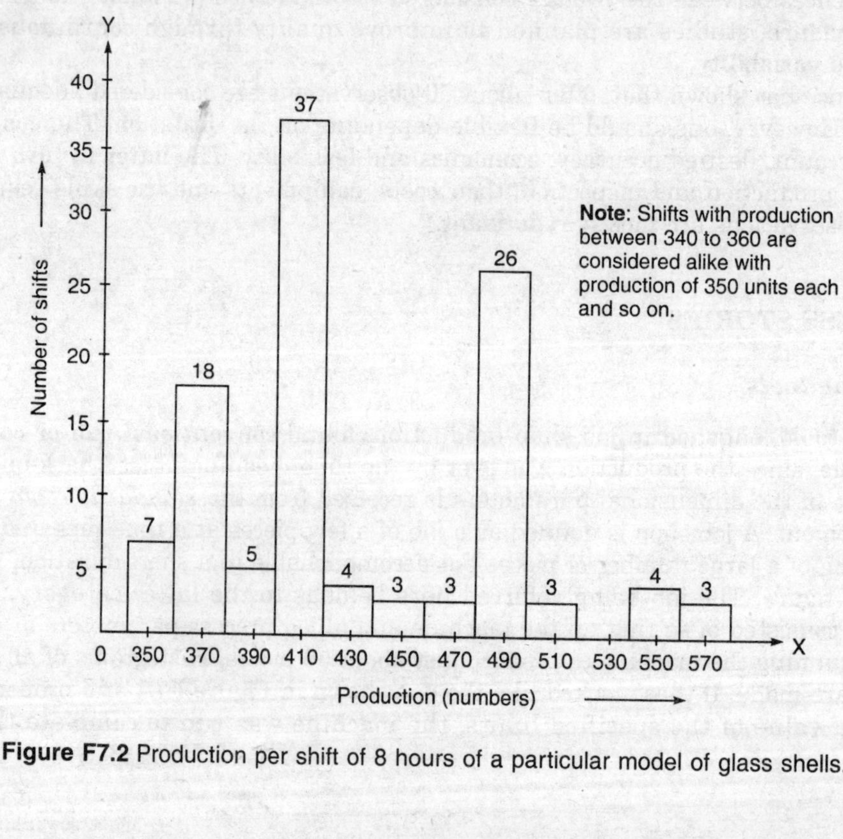

Figure F7.2 Production per shift of 8 hours of a particular model of glass shells.

only about one fourth of the total production cost. All other costs are fixed and per unit cost comes down with the increase of production. If production is enhanced from the present level of say 430 to new level of say 555 (an increase of about 30 percent); the production cost reduces to below 85 percent. It is considered possible if favourable levels of causative factors that prevailed during the period of freak beneficial period are traced and sustained. This calls for study of histogram of various likely causative process parameters like temperature and viscosity of melt glass as also scatter diagrams of these factors versus production. Likely recurring gain of 15 percent in the cost of production sets guide lines for efforts and investment, in remedial measures. Slackness on the part of any one member of the 12 member team or one's absence even for short intervals has substantial adverse effect on the production of the team. This in turn hurts the interest of every member of the team in terms of loss of production incentive and also reduces the return to the company, besides likely adverse impact on its competitive potential.

A provision for extra suitable person, capable in all operations, to replace the team members in rotation or in exigency, besides control of process parameters can completely achieve the mission of augmenting production, productivity and profitability.

It is absolutely necessary to assess the potential gains versus extra cost involved, if any, in the remedial measures and confirmation of these through trial runs before permanent implementation. Such reviews should be repeated for further identification and exploitation of potential for improvement. Similar situation was experienced in a large cycle manufacturing plant too.

(c) Chemical plant

Now look at four histograms shown in Figure F7.3. The difference in production of a chemical between 239th and 240th days, as also between 258th and 259th days (variation between two consecutive days or short term variation) are negligible. The causes for these minor deviations are difficult to locate for elimination. However, the collective pictures of 258th and 259th days

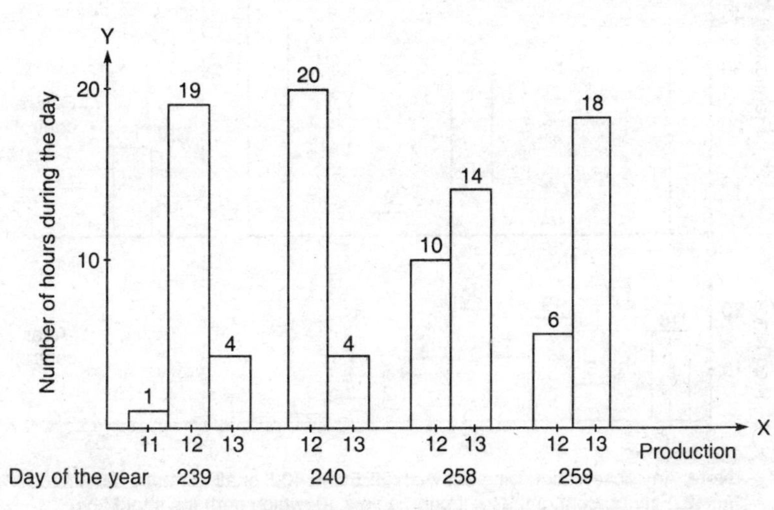

Figure F7.3 Histograms showing frequency of *production of a chemical per hour in 00's of litres,* for a pair of two consecutive days each.

together versus that of 239th and 240th days shows unambiguously a favourable change, a deviation observed over a longer period or interval of time called long term variation. The potential for increase of 1200 litres per day or about 4 percent is visible, once again at no extra cost except that of raw material. There is saving on energy, human resource and other heads. The cause, for this beneficial change, is considered relatively easy to trace and perpetuate. Another aspect that can be noticed is that the production figures are recorded in multiples of 100 litres, due to constraints of observing facilities. It will do well to record the production figures either, in smaller *least count* or by *shift* instead of by an *hour* to get better picture of variations. It is likely to be more useful in investigations.

Such situations are often encountered in manufacturing and service organizations and provide enormous opportunity, ready to be harnessed.

(d) Heavy electrical

A heavy electrical equipment plant was engaged in drilling about 4000 holes for tubes in condenser plate, with tolerance of 40 to 52 in coded units. The percent non-conforming holes varied from under one to almost hundred. Sometimes all over size, some other times all under size. Some times even nil non-conformity too was observed. This fact by itself is indicative of the process being capable. The histogram, shown in the upper half of Figure F7.4 presents frequency of one hundred holes at random. This is suggestive of *high* average diameter and a capable process. Generally, the operators have tendency to play safe and set for under size to avoid over size holes, which are more difficult and expensive to rectify. The over size holes need to be welded first with steel of the same alloy, before re-drilling to size.

We also observe, concentration of values at 45, 50, and 55 ; all multiples of five. The measuring instrument used had a least count of 5. This is too large in comparison with the

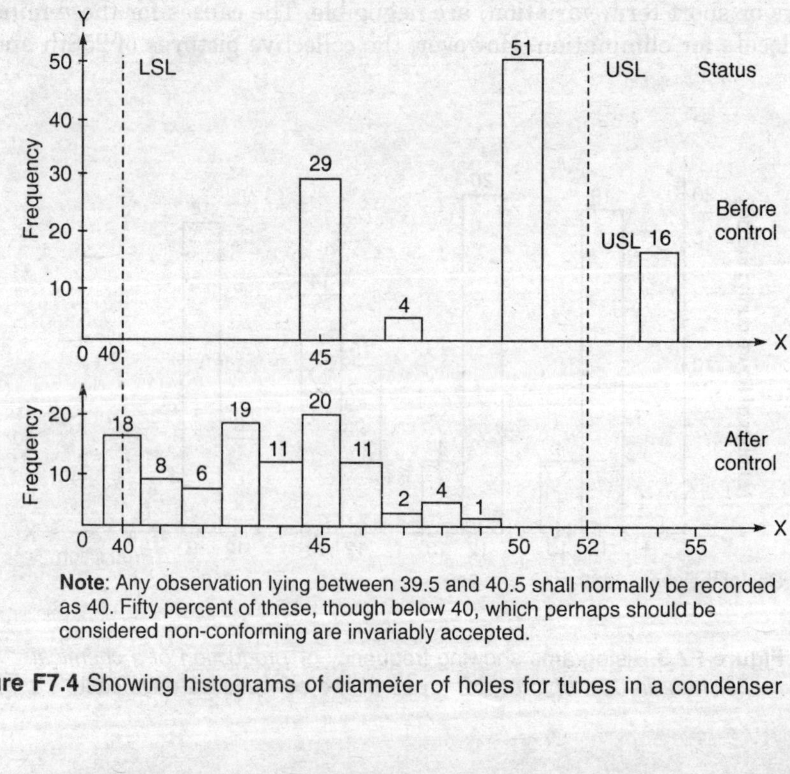

Note: Any observation lying between 39.5 and 40.5 shall normally be recorded as 40. Fifty percent of these, though below 40, which perhaps should be considered non-conforming are invariably accepted.

Figure F7.4 Showing histograms of diameter of holes for tubes in a condenser plate.

tolerance range of only 12. An instrument with least count of '1' was recommended. In the meantime, process control was initiated, with advice to the operator to make eye estimate of the size up to least count of one, even while using the same instrument, till new one is procured. The histogram of one hundred holes of next plate is shown in the lower half of the Figure F7.4. It certainly shows distinct improvement. However, now bias in favour of undersize is obvious and concentration of values at multiple of 5 persists, though to a lesser degree.

(e) Plastic textile

In a plastic textile mill the diameter of wound bobbins was checked for conformance to specifications with the help of go/no-go gauges. The number of non-conforming bobbins in samples of 20 bobbins each, varied from 2 to 12. Surprisingly, the number was always even, a very unusual and unexpected phenomena. It is very difficult to explain this unintended bias on the part of all associated with the inspection and segregation. The differences between the observed phenomena and expected natural phenomena (frequencies) are substantial. It gives undisputed evidence on the feasible scope of improving the validity of inspection results and hence a valuable support for subsequent efforts in improving conformity. See Figure F7.5.

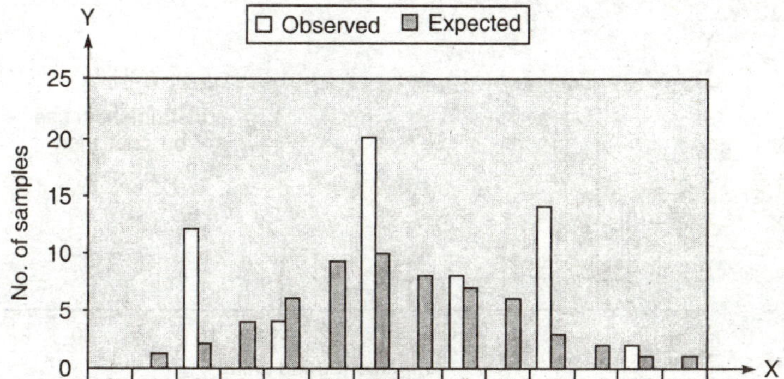

	0	1	2	3	4	5	6	7	8	9	10	11	12	13
Observed	0	0	12	0	4	0	20	0	8	0	14	0	2	0
Expected	0	1	2	4	6	9	10	8	7	6	3	2	1	1

Number of non-conforming bobbins

Figure F7.5 Histogram (It is called Bar Chart, when applied to attribute data) showing frequency of non-conforming bobbins in samples of size 20 each.

(f) Tele printers

Now, let us proceed to look at Figure F7.6 showing histograms of inspection results of the same 25 units on two sets of measuring electrical instruments, one each used by production and inspection personnel. The glaring difference, between the two measuring instruments, arising from the bias, is beyond any shadow of doubt. Since, identity number of the units measured on each instrument had been maintained, the pattern of the scatter of these observations will reaffirm the conclusion arrived at.

The consequences of this bias, can be appreciated, if the acceptable ranges were 1 to 10 or 6 to 15 or 11 to 20. Even though, the process is capable, the non-conformance might vary from zero to one hundred percent, causing all kinds of chaos in the minds of all concerned and on the shop floor too. The visible multi-peaks suggest that it is possible to produce components in the narrower tolerance range of 5. To assess further scope, more evidence needs to be gathered.

Unless, such behaviours are understood, the need for any remedial measure is never felt and hence the situation continues by default. These simple analytical reviews bring out the hidden truths, reveal the potential for improvement and provide vital clues to pursue for the solutions. The efforts in these directions take the organization closer to the goals of ppm or six sigma or zero non-conformity.

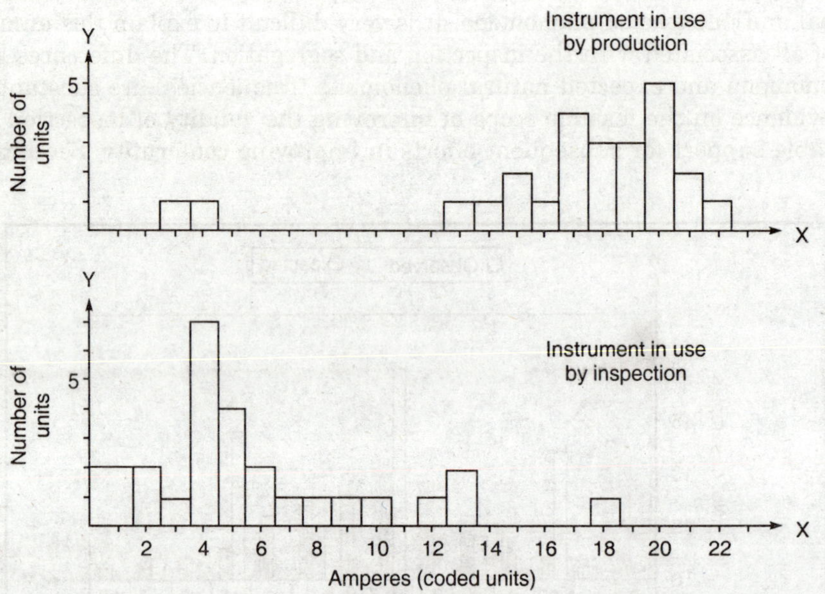

Figure F7.6 Showing histograms of measurements of a parameter (Amperes) of 25 components of a Tele printer, on two instruments, one each in use by production and inspection departments.

(g) Shaving blades

Let us look at Figure F7.7. It shows *honing angle* of 50 *shaving blades*. It is seen to vary between 18 and 21 degrees. Since, any value between 17.5 to 18.5 degrees will be read and recorded as 18 degrees and so on; the range of variation or dispersion may be considered to be (21.5 – 17.5) = 4.0 degrees, against the specification of (25.0 – 15.0) = 10.0 degrees. Process capability index is defined as the ratio of the tolerance range to the corresponding process capability. As stated earlier, the process capability is estimated as maximum variation (range) observed in a fairly large sample (50 is considered good enough for most practical situations) when the process is under the state of statistical control or the influence of chance causes only. In this case the process capability is estimated as 10/4 = 2.5 or 250 percent. The process is really very good. It has the potential to surpass six sigma status if the settings affecting honing angle are not ignored too much. The shifts, in the average, between 18 and 22 degrees, in such situations, shall be considered harmless by most in the industry. However, Dr. G Taguchi's concept of loss function does not recognize this as an excellent situation. He defines social loss equivalent to 'k' times the square of the deviation from the target, the middle value of the specification or the optimal value that delivers the highest performance. In this case, the target is 20 degrees. Assuming that the honing angle of 15 or 20 degrees renders the blade as scrap, the loss is its sale price, say Rupee one only. The present price of one blade is about Rupees seven only. The deviation at this extreme is (25 – 20) = 5. Its square is 25. Thus 25 k = 100 paise. Or k = 4. This implies that a blade with honing angle as good as 19.5 shall imply a social loss of 4 multiplied by the square of its deviation, which is equal to 20.0 – 19.5 = 0.5. This equals 4 (0.25) = 1 paise each. Like wise one with honing angle 16 or 24 degrees shall imply a loss of 64 paise per blade, implying that the real worth of the blade to the user in terms of number of smooth shaves is only 36 paise. In the above example the honing angle is estimated as 19.5 degrees. It may noted that the distribution is symmetrical around its mean and is considered satisfactory.

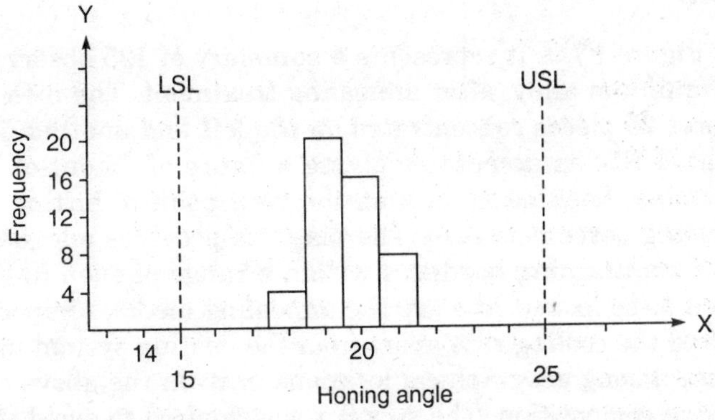

Figure F7.7 Histogram of honing angle of shaving blades.

Now let us consider another parameter of blade namely *centrality*. This has only upper specification of 0.10. The histogram is shown in Figure F7.8. The shape of the distribution is not expected to be symmetrical or normal. It is expected to be skewed to the right. About one fourth of the blades with centrality between 0.05 and 0.07 are considered avoidable through process control. Thus the process seems to be capable of producing blades with centrality within 0.04 which is 40 percent of the maximum specified. This process is also very good. The process capability index once again is 2.5.

It is worth recording that the production department wanted to import new capable machines to manufacture blades to conform to desired specifications. New machines may be needed for augmenting production, but certainly for not enhancing conformance. The present machines are in excellent condition.

In another factory, making automotive components, the production chief, made a similar demand for grinding machines. He was asked to assess the process capability with the above procedure. He was assured that in the of event of these being found incapable, necessary funds for procuring new machines shall be provided, since the management is interested in marketing only quality product. The study confirmed that the machines were capable and were in good shape. The production chief satisfied himself that by using statistical process control charts, satisfactory product was obtainable and withdrew his demand for new machines.

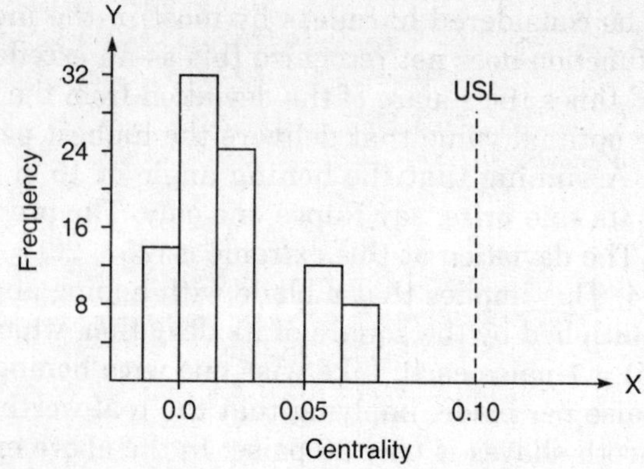

Figure 7.8 Histogram of Centrality of shaving blades.

(h) Die casting unit

Next, let us look at Figure F7.9. It represents a summary of 125 observations on hardness of die cast pieces of aluminium alloy, after annealing treatment. The men in the industry, who matter, shall interpret 25 pieces concentrated on the left and another 50 on the border near upper specified limit (USL) as acceptable giving a figure of 75 out of 125 or 60 percent as acceptable or conforming. Looking at it, from the view point of hitting the target,one might say that the conforming percent is zero. The diagram provides adequate evidence that the process is capable of maintaining hardness within a range of even half the specified range. The factors that need to be looked into are the annealing media, temperature, composition of the alloy, duration and the cooling rate apart from the heating system in the furnace and any systematic differences among geographical locations in it. In the above case, it was mix up of two batches of different composition. The supplier was advised to avoid this. He complied with the advice. His cooperation solved or eliminated the problem.

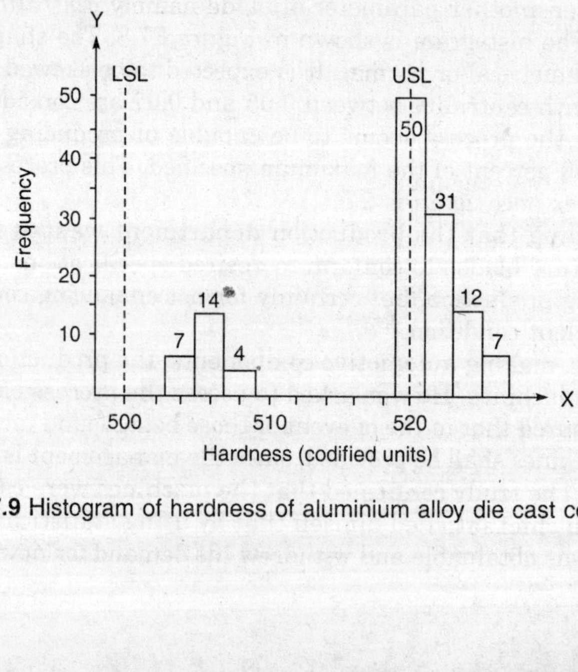

Figure F7.9 Histogram of hardness of aluminium alloy die cast component.

A picture similar to Figure F7.9 emerged when resistance expressed in ohms of elements, called tracks used in the assembly of *potentiometer* was summarized in the form of histogram.

(i) Engineering—Automotive parts

One commonly comes across patterns of histograms similar to those shown for two batches A and B in Figures F7.10. The reasons for these are traceable to biases among inspectors and or measuring devices. Besides, there is tendency to pass border line cases close to specified tolerance boundaries, even when there is abundant evidence that the processes are capable and these are avoidable through process control. Such components that are within permissible tolerance limits but do not conform to specified mean and normal pattern do create problems during assembly.

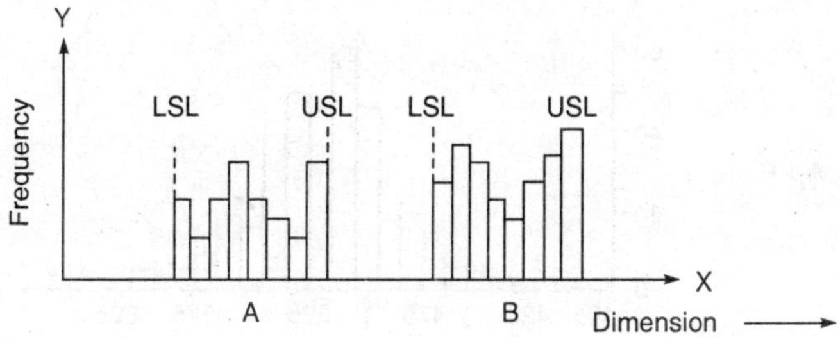

Figure F7.10 Histogram of a parameter of a component for two batches.

(j) Plastic fibre

Now let us look at a set of histograms of breaking load of plastic fibre, shown in Figure F7.11. The one at the bottom is the composite picture of production of a plastic fibre run on two machines, extruders,in this case, A and B for two weeks I and II. It has four or more peaks. It is not uncommon to come across histograms which are flat called rectangular. These may have many peaks or modes. This is a clear sign of shifts in the process settings giving rise to changes in the means. This becomes more clear, if we split the histogram into four histograms, one each for each machine and week. The highest peaks of these are respectively at brcak load of 500, 510, 540 and 580. Even, each one of these histograms has multiple peaks. A look on the right side of top two histograms for extruder 'A' shows gradual decline in its frequency. If this pattern were repeated on the left side, a crude estimate of process capability may be assessed as 90. May be, further studies, shall provide a better (smaller or narrower) estimate. In order to avoid non-conformities of breaking load less than 498 (or say, 500); the process mean should be aimed at (500 + 90/2) = 545 or say 550. The parameters that need to be examined or taken care of, in this situation are temperature, speed and tension.

(k) Eelectroplating of automotive and allied components

The shapes of histograms might differ with the choice of grouping or the width of the class interval, represented by a bar. Unintended wrong interpretations are possible for lack of adequate knowledge and or experience. Figure F7.12 shows three histograms of same 54 observations, made on the strength of etching bath in a plating shop (of rings used in assembly of automobile engines) with grouping (class width) of 4, 8 and 12 coded units

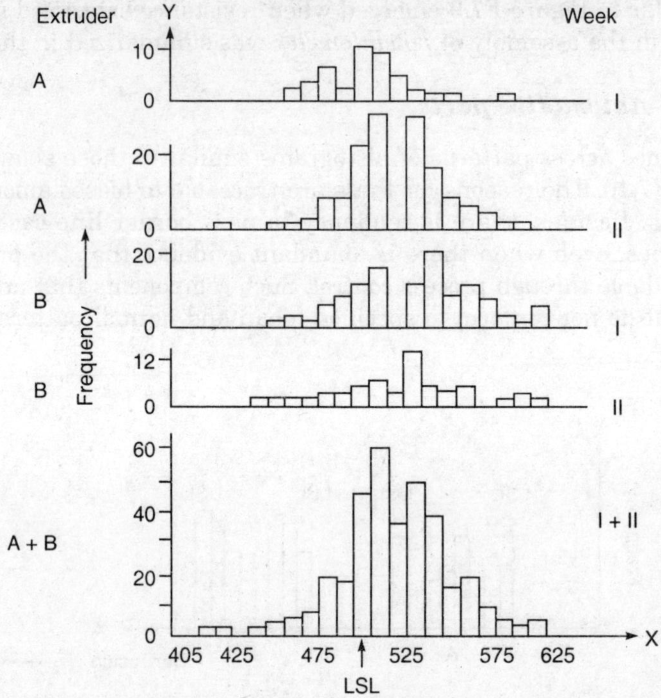

Figure F7.11 Histograms of breaking load of plastic fibre with minimum breaking load of 498 in coded units.

respectively. The histogram shown at the bottom, clubs observations between 154 to 158, 158 to 162 and so on. The middle histogram clubs the observations between 154 to 162, 162 to 170 and so on. The histogram at the top, displays the same data by forming groups from 154 to 166, 166 to 178 and so on. The histogram at the bottom shows largest number of peaks with a feeling of likely capable process. The one at the middle shows with not enough confidence on capability of the process. While the one at the top, just shows a flat or a rectangular shape, indicating that the process has been left to itself with no central tendency and that the process variation is three times that of specified range or interval.

There is need to be cautious about divergent interpretations possible with different choices of class interval. The choice of class interval should be judicious. Never the less, all the three shapes do indicate potential for improvement. The thumb rule is that the number of class intervals should be equal to the square root of the number of observations available.

The tolerances for plated components of cycles are wider while that of printed circuit boards are narrower than the one's, just discussed. In all these cases, adequate attention is not paid to maintain concentration of various baths within specified norms. Surprisingly, in all the above three situations the observed variation is three fold of their respective specifications. Improved maintenance of the batches through guided replenishments and replacements had salutary effect on percent non-conformities. Simultaneous control of other bath parameters like current levels brings further improvements.

Such rewarding results were also obtained in cell house, producing hydrogen and oxygen, in vanaspati manufacturing plants. Similar, was also the experience in the plating process of wires used in the manufacture of telecommunication cables.

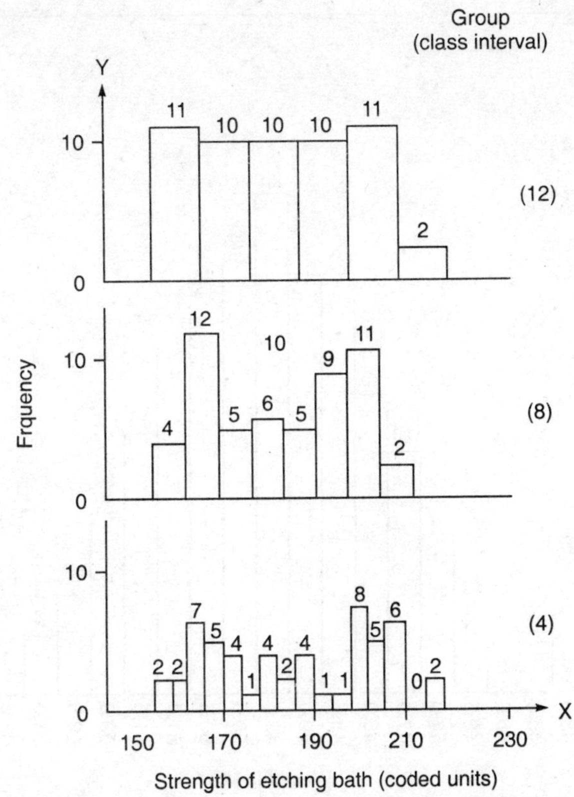

Figure F7.12 Strength of etching bath, plating shop—Specifications 180 to 200 (coded units).

(l) Fertilisers

A most interesting fallacious use of histogram came to light in packing section of a large fertilizer plant. A histogram of weight of 200 bags picked up at random from a day's output of all machines showed a stable symmetrical shape like that in Figure F7.13. It was concluded that the process was in a state of statistical control and the observed variation was more than twice the specified tolerance range, some thing needs to be done to reduce the same. The cause of excess variation, propounded by the technical experts was the presence of the dust in the packing room interfering with the performance of sophisticated filling and packing machines. Hence air-conditioning of the huge filling and packing hall was recommended, little realizing that the dust was emanating from the fertilizer itself. The ground reality of the situation was that the settings or adjustments of various nozzles of different machines, were not receiving due attention, for lack of conclusive information. The histograms drawn separately for these sources confirmed the capability of the existing facilities. Air-conditioning was avoided. Appropriate frequent dusting and adjustment of settings for standard weight yielded satisfactory results.

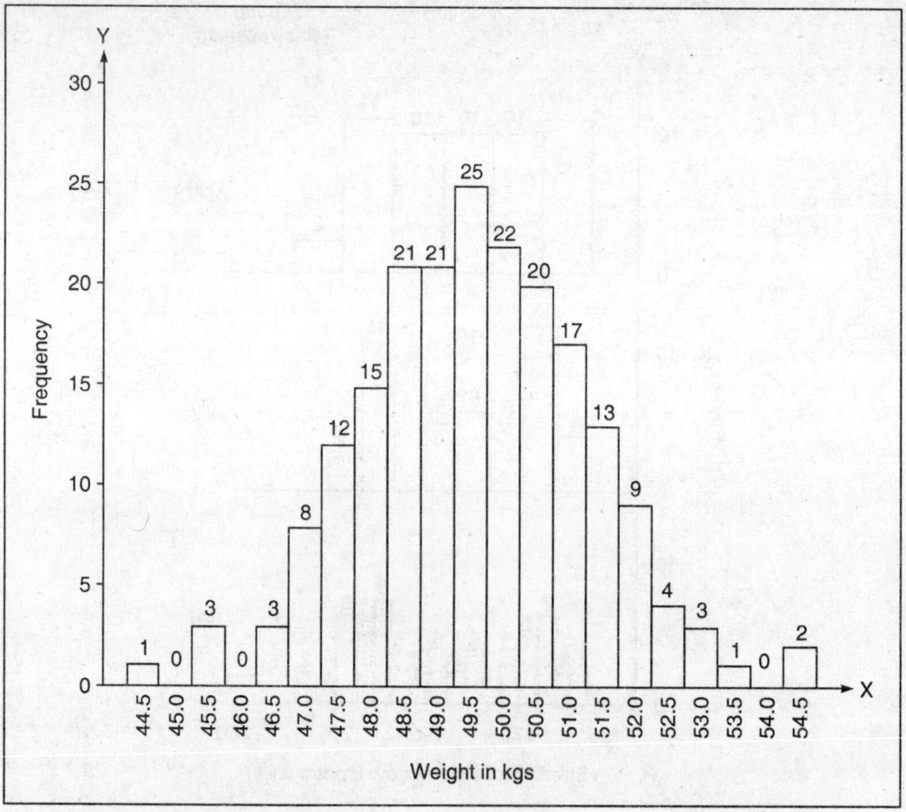

Figure F7.13 Histogram of weights of 200 bags of fertilizer (in Kilograms).

(m) Conclusion

Histogram is yet another simple tool that can be practiced, conveniently to review the performance of a process for making improvements. All it requires is a pencil or a pen and a paper, in addition to existing facilities. It has one disadvantage. The summary of the data presented in the form of the histogram ignores the sequence of production. Thus some vital clues; of process imperfections, possible from the examination of the pattern of sequence; get lost. These get made up by the use of run charts, the subject matter of next chapter. The use of histograms compensates by highlighting, even small deviations from the target prominently through review over convenient, logical and meaningful periods. The examples sited above have thrown open an enormous potential for its exploitation with benefits that one cannot afford to sacrifice.

The preceding, three tools, namely, stratification, scatter diagram and histogram play a vital role in resolving the problems. These tools, provide very vital clues that help the investigations to trace the root cause as also their remedies. There might arise some situations when the problem does not get resolved and generate a feeling of failure. It would be unfair to accept that status. The findings of the study, have taken us thus far, to reveal one way that does not take us to our destination and provide leads and direction for our next steps to reach our goal.

8

RUN CHART

8.1 DEFINITION

A Run Chart may be defined as a graphical presentation of data on a chart with sequence of production on the x-axis and the parameter of interest on the y-axis.

8.2 CHOICE OF SCALE

The time interval on x-axis may be as large a unit as 'a million years' as applicable in geological studies devoted to science of earth's crust, its strata and their relations or changes or study of fossils—things preserved in strata of earth or anthropology, the science of man. Alternatively the unit may be a century, as commonly referred in history; or a decade as in the case of census; or a year as in the case of usual annual financial reports of commercial organizations. On the contrary, the unit may be as small as an hour or a minute or even a second depending on the production rate or the speed with which the parameter under observation undergoes a change of a magnitude that causes concern to the observer.

The scale on y axis should match the needs of visual appeal or perception for meaningful interpretation and convenient for plotting depending on the unit or the least count of the observations available.

8.3 UTILITY

To make it a useful tool for understanding the bahaviour of the process, namely the resulting mean and the spread, it is desirable to indicate the desired target or the mean and the permissible upper and or lower tolerances viz the acceptable boundaries.

The trueness and precision of the observations or the measurement should be good enough for the purpose. The least count should be one tenth of the specified tolerance interval or the natural spread of the process whichever is smaller. A smaller least count may add to the cost while a larger may defeat the purpose. The repeatability and reproducibility errors should not exceed one sixth of the natural spread of the process or the tolerance interval for similar reasons.

Likewise, in the case of attribute data, expressed in numbers or percent should be recorded to the accuracy of approximately one tenth of acceptable or observed quality level whichever is better.

The interval for making observations, may be about half of the stable duration of the process; the spread, during which interval, is considered harmless. However, if the situation demands as a part of special in depth study or otherwise, the observations on all items produced may be plotted in the sequence of production, over the period of interest.

The success stories documented in (8.6) shall substantiate the benefits of satisfying the above requirements.

8.4 PRE-REQUISITES

It is pertinent to emphasize even at the cost of repetition, that the data need to be relevant, authentic and adequate; save in exceptional cases when one may have to be contended with a few observations or necessarily deal with large data or whatever is available. The chart will be able to portray only as much information as contained in the data.

Usually, about 25 to 30 observations are good enough to familiarize with the behaviour of the process. The period of observations should be wide enough to provide opportunity to all possible chance causes or sources, over which one has no control, to have influenced the process. These charts help to identify the periods of stable and unstable behaviours of the process, as also working levels of means during these periods. Such periods are then used to trace the sources thereof, from the relevant process data, available from the documented check sheets, specially designed to take into consideration the factors included in the corresponding cause and effect diagram. The reasons for unacceptable pattern(s) or deviation(s) need to be gone into and measures to avoid their recurrence are found and implemented to make and sustain necessary improvements. The efficiency is judged by the speed of detection of undesirable change(s), stoppage of process and its resumption to original or desired status. The effectiveness is considerably enhanced if all the perceptible *changes* occurring in the process are recorded faithfully as *remarks* or *foot notes*. The *changes* may include details like change of operator, tool, inspector, gauge, voltage, die, jig, fixture, temperature, lubricant, viscosity. Also the details of the nature and quantum of corrective actions taken, need to be documented like wise. These together help in developing and updating standard *work instructions* that aim at minimizing the variability and sustaining it.

8.5 EXTRA BENEFIT

This is a very important and useful tool in sustaining gains actualized by identifying and solving a quality problem for improvement with the help of the six tools discussed hitherto. It must, however be borne in mind that a judicious use of this tool can often result in *break through* solution to a *chronic* problem, even though it is *primarily* intended to solve *sporadic* ones.

8.6 HARNESSING CHARTS FOR IMPROVEMENT

8.6.1 Attribute

(a) Engineering industry—press shop—drawing operation

The process consists of drawing a circular disc, cut from a tin plate, into a hemispherical bowl. This bowl forms a part of oil tank of an hurricane lantern. About ten percent of the pieces developed crack during the operation of the drawing. The cracked pieces were scrapped. A sample of ten consecutive operations was observed every hour and number of cracked pieces was recorded. Twenty five such samples spread over three shifts were observed. As a thumb rule, the sample size, expected (average in the long run) to contain one non-conforming unit each is considered appropriate and twenty five such samples at suitable intervals are considered adequate for the purpose. The data are shown in the form of a run chart in Figure F8.1.

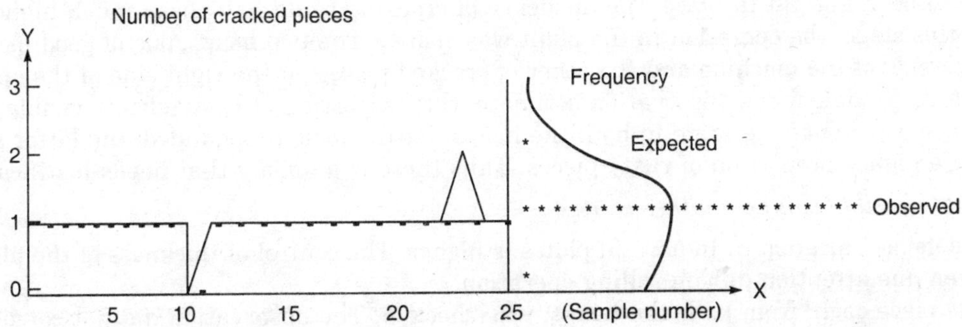

Figure F8.1 Number of cracked pieces observed in 25 samples of ten each.

It is observed that:

There are two runs of ten (one to ten) and eleven (twelve to twenty two) points (samples) respectively, viz with one crack piece each only.

Except for 11th and 23rd samples which had zero and two cracked pieces each, all the remaining 23 samples had one cracked piece each. The knowledge of binomial probability tells us that if samples of size ten each are picked up at random from a box containing large number of beads (90 percent green symbolizing good pieces and 10 percent red symbolizing cracked pieces) repeatedly *twenty five* times; then on an average in the long run; 9, 10, 5 and 1 samples respectively will show 0, 1, 2 and more than 2 red beads respectively. The bars and the curve, shown in the right portion of figure F8.1, compare the observed and expected situations.

Note:- A Chi-square test to examine the proximity or otherwise of observed and expected frequencies shows this departure to be much wider than that can be expected under similar situations.

Both the above observed phenomena are considered non-random or un-natural. This implies that there is strong possibility of some special or systematic cause behind this consistency (precisely, one piece, out of every ten consecutive pieces, produced in a small span of time, getting cracked regularly), which is too good to be true. The visual display makes this, otherwise obvious inference, convincing.

Yet, most people, consider this situation as satisfactory or normal stable behaviour, expected of a process prone to ten percent non-conformance, out of ignorance. This creates complacency and a golden opportunity, to look for improvement, is lost.

This led to observation of consecutive operations more minutely. It was observed that whenever a piece cracked, the operator would clean the die with a greasy cloth before re-starting. Subsequent to this, every 13th or 14th piece was getting cracked. This was observed three times. To confirm this periodicity, the die was cleaned after every ten operations and lo behold, all the thirty pieces were found to be free of any crack whatsoever.

A discussion with the chief engineer of the plant followed. It ran as follows:

What is (are) the cause(s) for the incidence of cracking?

The Indian supplier, of tin plates, does not take adequate care in packing and storing. The tin plates accumulate a lot of dust and also become rusty. These cause the sheets to crack.

In support of his argument, he showed additional operations introduced by him, to remove the dust and the rust. He further added that the incidence was only one percent when imported plates are used, without these additional processes. No doubt, had these extra measures been not put in place, the incidence of cracking could have been much higher.

At this stage, the operation in the plant was visited. The two heaps, one of good pieces on the back side of the machine and the other of cracked pieces on the right side of the operator were seen. By naked eye observation estimate, the proportion of rusty sheets visible to the naked eye was about the same in both the heaps. By the logic propounded, the latter should contain a higher proportion of rusty pieces. Thus there is anomaly that needs justification.

What possibly can be another cause?

The gauge variation in Indian tin plates is higher. The control of thickness of the plates is not given due attention during rolling operation.

One piece each, from both the heaps, was checked. The observation did not corroborate the above argument.

The chief engineer was again shown, the recurring behaviour of cracking when the die is cleaned after the incidence of cracking and its absence when the die is cleaned in anticipation after every ten operations.

The above leads to the following conclusion:

- Without contradicting the claims and the facts reported by the chief engineer, it needs to be understood that an Indian tin plate may require cleaning of the die after every ten operations and dismantling and re-assembling of the die once a day for crack free operation.
- While processing imported tin plates, these intervals may be 100 operations and a week respectively.

Thus a status of ten percent non-conformance can and has been changed to a status of parts per million (ppm) or six sigma status, yielding over ten percent additional production at no extra cost giving tremendous boost to profitability accruing from this operation.

By way of information, it may be added that:

- The tin plates were in short supply and its distribution was regularized through quota system at controlled price.
- Extra over ten percent more production of hurricane lanterns was possible.
- The scrap was in fact sold at higher price to manufacturers of products of smaller sizes.

What shall the scrap manager do?

(b) Engineering industry—Job shop—welding process

Number of non-conformities per ten meters of welding jobs, done during the day for a group of eleven workers were recorded for two weeks - twelve working days. This number was used as an index of performance of quality. The wisdom of clubbing variety of jobs performed by several workers is questionable. However it may be inevitable and not undesirable either; if it is a team work on huge jobs, the output of each worker is relatively small and their capability on the variety of jobs is almost alike. Never the less, the index is monitored in the form of a run chart, shown in Figure F8.2. It sounds a signal if the performance is too good or too bad in relation to the past acceptable norms. It is seen that the incidence of 7.8 nonconformities per 10 meters on the fifth day is conclusively not consistent with the average of 2.5. This sporadic event signals adverse change and calls for search and implementation of corrective measures. It is unlikely to be worker dominant, since even the average of all the workers has shot up. Its root cause is more likely to arise from system failure, such as; use of wrong, badly preserved or inadequately heated electrode; low voltage of current supply. The system belongs to the management.

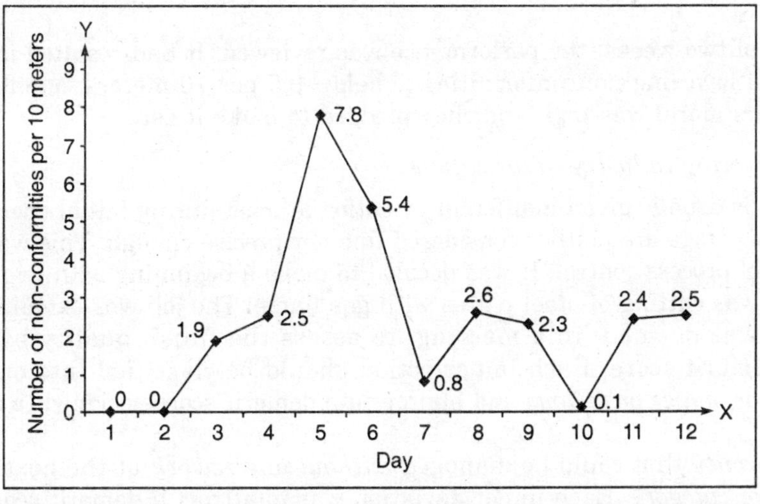

Figure F8.2 Run chart showing welding nonconformities per ten meters per day.

The company luckily had the habit of recording findings of the inspection with pertinent details. These details were used to make a two way table called a *matrix table,* as shown in Table T8.1. This shows the nature or type of nonconformity and its likely source as hunched by the professional inspector.

Table T8.1 Showing incidence of nonconformities stratified by its type and likely source

Source→ *Type ↓*	*Current setting*	*Electrode*	*More gap*	*Gas cold*	*Total*
0	*1*	*2*	*3*	*4*	*5*
Under cut	10	0	5	0	15
Porosity	8	4	0	2	14
Blow holes	0	3	0	1	4
Cracks	0	0	0	0	0
Total	18	7	5	3	33

The nonconformities were also stratified by operators. It varied from ZERO per ten meters for the best performance to EIGHT for the worst. The performance of the remaining nine operators using the same index were 0.5, 0.7, 1.0, 1.8, 2.1, 3.1, 5.2 and 6.2.

Equipped with these facts, a meeting was held with all the welding operators in the presence of concerned technical support personnel. The status of 2.5 non-conformities per 10 meters as also wide dispersions were brought to their knowledge to seek advice for improvement to reduce avoidable tangible and intangible losses. They expressed dissatisfaction with the current status and vouched potential for doing better. They were encouraged to offer suggestions. They asked for *bigger grinders* and better maintenance of welding equipments, particularly their *current setting*. The former was not considered logical but conceded to give them a chance to demonstrate. The latter was accepted on the basis of the data of Table T8.1 and steps taken. It was however informed that inspite of poor maintenance, some of the operators were getting it right by starting work only after getting these set right. Such dedicated and committed operators were complimented and encouraged to train others.

At the end of two weeks, the performance was reviewed. It had resulted in breakthrough improvement of lowering nonconformities to below 0.5 per 10 meters, one fifth of the past figure. Operators moral was high and they pledged to make it zero.

(c) Heavy engineering industry—fabrication

Fabrication is usually given indifferent attention at least during initial operations, because these jobs at this stage are neither considered fine nor precise enough. This was a virgin area for application of process control. It was decided to make a beginning with two operators. The first operation was cutting of steel plates with gas flame. The job was examined visually for its fitness. It was decided, in a meeting, to assess the finish quality by assigning an appropriate demerit score. Each imperfection should be classified into one of the three categories, *minor, major or serious* and appropriate demerit score assigned to each plate that has been cut.

A *nonconformity* that could be managed *without* any *rework* at the next operation was defined as *minor category*. Each minor deviation was to attract a demerit score of *one*.

A nonconformity that could be considered *acceptable* after *rework* at the next operation was agreed to be classified as *major category* and assigned a demerit score of *three*.

A nonconformity that is not considered worthy of any repair whatsoever and is deemed as *scrap* was to be categorized a *serious* and assigned a demerit score of six.

Obviously a plate free of any nonconformity would attract a score of zero.

A *run chart* for demerit score for each plate that is cut was initiated for two of the operators designated A and B. The data thus obtained for four plates for operator A are shown at the bottom of Figure F8.3. However the demerit scores for both of them are graphically presented in Figure F8.3.

The status was discussed among all the five operators of the work area in the presence of their supervisors and executives. All of them recognized the objective evaluation for comparison of the performance offered by this procedure. They also visualized the scope for improvement and control that the demerit run chart would provide. The system was extended to all the five of them there after. The data for operator A and chart for all the five of them for next three jobs each are shown in the same Figure F8.3. The performance that started with a demerit score of over thirty converged to a demerit score of under ten per plate. With awareness aroused, it is considered possible to actualize further improvements. This gives a signal that any quality parameter that gets quantified becomes amenable to control.

It may be noted that *serious* nonconformity was never observed. It may be desirable to improve upon the criteria of classification and scale of demerit score for the imperfections in the gas cut plates, to make it more useful.

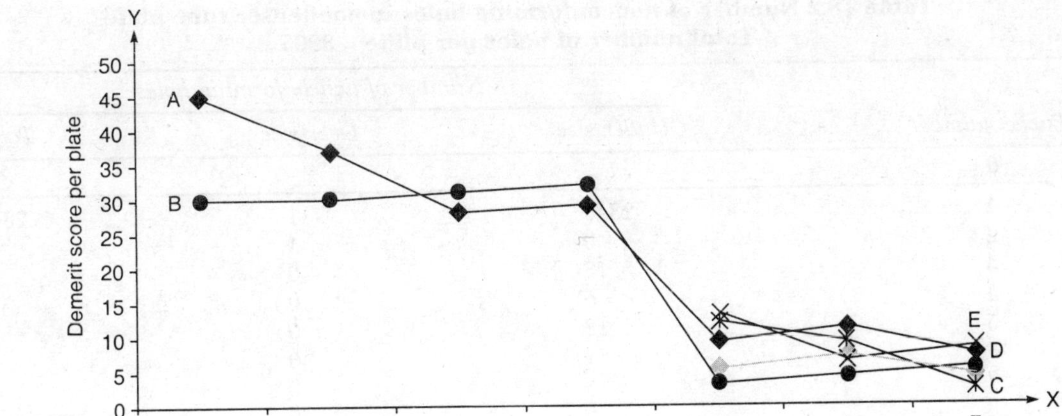

Plate No.	1	2	3	4	5	6	7
Operator A							
Serious (6)	0	0	0	0	0	0	0
Major (3)	14	11	8	8	2	2	1
Minor (1)	3	4	4	5	3	5	4
Demerit score	45	37	28	29	9	11	7

Figure F8.3 Run chart of demerit score of nonconformities observed on a gas cut plate.

(d) Heavy electrical—condenser plate drilling of holes

A particular model of condenser plate was designed to have 3905 holes each. The specified limits for the diameter of the circular hole were 30.40 + 0 .00 and 30.40 + 0.12 in coded units. After the job is done, all the holes are inspected with the help of go no go gauges of sizes 30.40 and 30.52 respectively by the authorized competent inspector. He marks the nonconforming holes for repair. The under size holes are drilled to size, while the oversize holes are first filled by welding with electrodes of same alloy and then re-machined. The repaired holes are inspected again for their conformance.

The rework in mending undersize and oversize holes was substantial and unpredictable, sometimes too little and on some other occasions too high. These caused hurtful delays and escalated costs. The past data for twenty one condenser plates was available. These are tabulated in Table T8.2

The same data are presented in the form of a run-chart in Figure F8.4.

A look at the above data shows that the number of nonconforming under size holes varied from zero to 3905 while that of over size from zero to 24 only. It goes to show that operators are cautious to avoid the latter because it entails more resources and operations. One plate consists of as many as 3905 or nearly 4000 holes. Another has 2344 nonconforming holes. Excluding these two the average number of nonconforming holes per plate are only about 12.4 or 0.32 percent only. With as little as only one nonconforming hole in 2nd and 6th plates out of about 4000 holes each, one need have no doubt about the capability of the process to yield zero nonconformity. The problem addressed with the aid of histogram has been discussed under the caption success stories in 7.5 (d) of chapter VII. Two separate run charts one each for under size and over size holes are likely to be more useful.

Table T8.2 Number of nonconforming holes in condenser tube plate
Total number of holes per plate = 3905

Packet number	*Number of nonconforming holes*		
	Under size	*Over size*	*Total*
0	*1*	*2*	*3*
1	2343	1	2344
2	0	1	1
3	13	0	13
4	5	0	5
5	13	0	13
6	1	0	1
7	3	0	3
8	5	0	5
9	3905	0	3905
10	6	17	23
11	13	24	37
12	8	0	8
13	2	0	2
14	3	0	3
15	0	21	21
16	4	0	4
17	6	4	10
18	26	0	26
19	20	0	20
20	24	0	24
21	16	0	16
Overall (Average)	305.52	3.24	308.76
Excluding 1st and 9th plates.	8.84	3.53	12.37

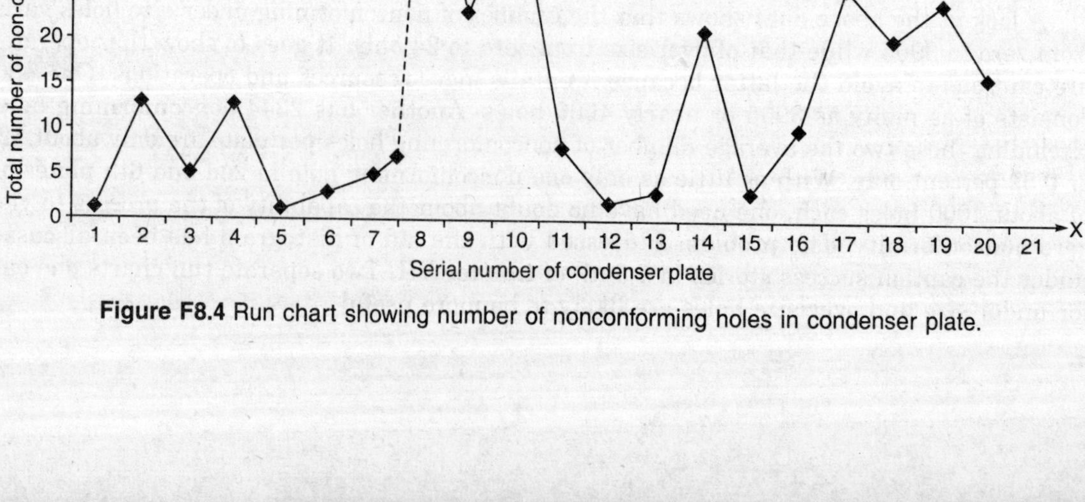

Figure F8.4 Run chart showing number of nonconforming holes in condenser plate.

(e) Engineering industry—job shop—machining operations

The shop was manufacturing turbine blades. The percent nonconformities was on the higher side of 30. This contributed substantially to rework in many forms, scrap and delays. A run chart on percent nonconformities was initiated and all pertinent details, such as machine or operation, nature of nonconformity and its likely source, were recorded. See Figure F8.5.

The number of components produced was small. Each component required machining of several dimensions and profiles. Several workers worked on several machines. The total number of nonconformities observed on all components produced during the day was divided by the total number of parameters checked on all the components, and the ratio multiplied by 100, to arrive at the desired figure of percent nonconformities. This clubbing looks irrational, but their was no alternative. The number of items, required to be processed, was small. The jobs changed very frequently. Plus point was that, a couple of in depth studies, wherever possible had demonstrated, the adequacy of the process capabilities. This encouraged a look at the cumulative picture to look for the causes of nonconformities and their preventive measures.

For quite sometime the status did not change for better. The actions as indicated by the nature of nonconformities was not taken for fear of hurting production.

The matter was discussed with the executives responsible for planning of allocation of jobs to machines and monitoring its progress. They were equally responsible for quality, but it was not given adequate attention. The reason was the delayed information on the previous day's performance. It reached them after the production for the day had commenced and any corrective measures will disturb the tempo of the production. The situation was remedied by summarizing and presenting the status, as obtained till about an hour before the close of the day. It is logical to assume that the on going processes will continue to behave in the same manner as hitherto. This enabled the executives to plan and implement the remedial measures before commencement of the production next day. The incidence of nonconformities declined to below five percent by the end of the first week and below one percent during the last week of the month.

The accompanying tangible and intangible benefits were tremendous. The absolute production went up three folds. With reduced non-conformities, the production of conforming product shot up four times. The delays got eliminated and deliveries advanced. It reduced controversies among human resource and enhanced mutual understanding to create healthy congenial environment worthy of emulation by others. The financial implications surpassed all expectations.

The above applications demonstrate the role of systematic studies in identifying the dominance of weak areas and also, whether these pertain to the domain of workers or management. The approach brings the two together in search of the solution to the problem on hand, rather than laying the blame on each other's doors. This cultivates harmonious environments and makes the target of ZERO NONCONFORMITY assailable.

8.6.2 Variable

Quality is universally acclaimed and accepted as *fitness for use*. The *degree* of Fitness for use as perceived and demanded by the user keeps on moving upwards, thus necessitating *continual improvement*. Absence of improvement is *fatal*.

Translating fitness for use to optimal product and process specifications entails detailed study, of all inputs; based on analysis of appropriate, adequate and true data.

CHART

Percent nonconformities observed in turbine blades

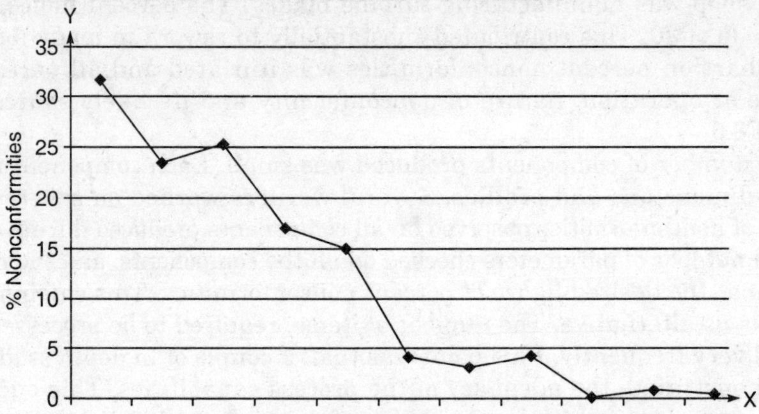

Day:		1	2	3	4	5	8	9	10	27	28	29
Data sheet:												
Production	No.	700	743	667	809	695	1040	881	1050	762	240	846
Non-	No.	222	174	169	138	104	44	26	44	0	2	3
Conformities	%	31.7	23.4	25.3	17.1	15.0	4.2	3.0	4.2	0.0	0.8	0.4

DATA SHEET

Classification by degree/ critcality of deviation:

Passed	No.	216	172	167	128	104	44	26	40	0	0	3
	%	30.8	23.1	25.0	15.8	15.0	4.2	3.0	3.8	0	0	0.4
Rework	No.	0	0	0	0	0	0	0	0	0	2	0
	%	0	0	0	0	0	0	0	0	0	0.0	0
Scrap	No.	6	2	2	10	0	0	0	4	0	0	0
	%	0.9	0.3	0.3	1.7	0	0	0	0.4	0	0	0

Source wise detail-number only:

Operator	15	75	166	84	99	40	26	34	0	2	3
Tool-fixture-jig	0	0	0	0	0	0	0	0	0	0	0
Cutter	86	31	3	2	0	4	0	0	0	0	0
Machine	0	0	0	0	0	0	0	10	0	0	0
Inspection	2	2	0	0	0	0	0	0	0	0	0
Previous stage	115	66	0	52	5	0	0	0	0	0	0

Source wise detail-number only:

Stage(s)	10-12	13-15	11-12	16-189		13-15	13-15	13-15	-	28	10
Operation(s)	1,2	1,10	3,8	1,112		18	18A	18	-	12	17
Machine(s)	36,48	49,50	45,46	35,49	35,3650		34	34	-	23	50

*Repeaters are indicative of areas needing priotity attention.

Figure F8.5 Master Control System (CHART + DATA SHEET)—Attribute.

For instance, consider *fitness for use* of detergent for washing garments. It implies its power to brightly clean the fabric at competitive price, preserve its colour and life, without in any way harming the hands or skin of the user on contact and emitting obnoxious gases that may be inhaled inadvertently. The user will indeed get delighted if in addition it benefits the skin and the gases emitted are not only pleasant but also useful to the body. It is, obligatory to: thoroughly examine all the properties of all ingredients used; determine their optimal mix and the formulation procedure including preservation and packing to ensure shelf life against environmental hazards. The optimal washing procedure also needs to be determined and communicated.

Consider yet an other example of *shaving blades*. The user demands more number of smooth shaves per Rupee. This needs to be translated into specifications for : the direct and indirect raw materials; equipment and facilities for production and testing; procedures for manufacturing, preservation, packing, shaving and disposal; finished product and documentation. Each of these needs to be dealt with in perfect and sufficient detail. For example, these shall include:

* Steel alloy, thickness, width, and strength.
* Sampling method, size, frequency and acceptance criteria.
* Capable equipments and their maintenance schedule.
* Competent human resource.
* Centrality and Honing angle of the finished blade.

The above list is not exhaustive!

In all such cases, the *optimal* level of each process and product parameter is unique, designated as the *target*. The *ideal quality* is achieved if each parameter is *dot* on the *target*. Any *deviation* from this *target* implies *less* degree of *perfection* necessary for *fitness for use*. This imperfection is quantified as *social loss* deemed proportional to the square of the *deviation*. It may be recalled that Mean Square Deviation is called *variance* and its Square Root as *standard deviation,* universally used as a measure of variation. Emphasis on *continual war* on *reduction* of *social loss* to society by Dr. Genichi Taguchi is synonym with the *optimal* approach to *hit* the *target* with *minimal variation* around it.

In some situations the loss or hurt caused by *poor quality* shall be many times more than the cost of the product. It might even prove *fatal*. Therefore such deviations need to be pegged at unavoidable *minimum* level in the situation on hand. This *minimum* keeps moving downwards with Cost Effective—Improvements in Process Knowledge and allied Technology. The *art*, of *translating customer's perception of fitness for use into compatible specifications* **for** *appropriate product* and *process parameters* and *taking all actions necessary* to *achieve these,* is indeed *putting quality function* in *place*. This is popularly known as *Quality Function Deployment (QFD)*

Often, there is satisfaction in meeting the specified tolerances in legal sense. This results in unintended complacency and sustains the culture of *stagnancy* with associated disastrous consequences. The leaders however do not remain contended and break the ice by making *improvements* in product and process designs through updating skills of human resource and allied systems with emphasis on augmenting *return on investment (R 0 I)*. Dr. J M Juran, the world *quality guru,* states the axiom that *all improvements are project by project only.* The sure way to *assure quality* is indeed to build it into the product. All one needs to do this, in most economical way, is to learn the *right* way and to *do it right first time*.

Let us see How *charts* help us in this *noble task?*

Prevalent Situation:

In most industries even today, one product parameter of an item is viewed at anyone point of time. Its literal or legal compliance to specifications including border line cases is interpreted as satisfactory situation. In good faith, the human resource complacently sustains this culture. The data thus generated form a part of history and adore the archive.

Desired Scenario:

It is desired to demonstrate that the following steps pave the way for alternate cycles of improvement (breakthrough) and control.

* Present the data in simple pictorial form consisting of run or appropriate control chart. [See IS : 397 parts 0, 1, 2, 3 and 4]
* Supplement the above by a histogram. [See IS:7200 parts 1, 2 and 3]
* Look for the short and long term patterns and trends of variation.
* Use the clues provided by these events as guide to investigate the sources of variation.

Cause and Effect Diagram and associated Critical Failure Mode and Effect Analysis arrived at a brain storming session among all concerned catalyse and sharpen this activity.

* Confirm the findings.
* Develop procedures to hit the target and to reduce the process variation around it.
* Implement and monitor the procedure so developed.

All these steps require *PASSION* for improvement and little investment, if any, to channelise the energies in the *RIGHT* path to become leader and reap accompanying benefits. Are there any options? Let us consider four examples. These represent Chemicals, Heavy Electrical and Automotive Components organisations.

(a) Fineness of a Chemical

In a chemical plant manufacturing fertiliser, pulverised rock fineness passing through 100 mesh screen, is an important parameter. The practice was to look at, one observation at a time, at the conclusion of a batch. This does not communicate anything except that by and large the attribute or the parameter conforms or does not conform to the specified tolerance(s). A glance at all the observations compiled in column 2 of Table T8.3 for a fortnight tells us about its Range, 4.0, the difference between the largest value (94.2) and the smallest value (90.2) encountered. All this sounds satisfactory in the light of specified minimum value of 90.0 and 'no' alarming signals are perceived to act for improvement. These data have been plotted in the form of a Run-Chart in the top of Figure F8.6. A careful look now highlights trends, sudden jumps, stable and unstable periods, leading to provocation that if reasons for such deviations or behaviour arising from avoidable sources are traced and acted upon, the product will be more uniform and hence superior resulting in better customer satisfaction and associated rewards.

Batch S. No.	MFG. Date Mar.	X Fineness	R Moving Range	Batch S. No.	MFG. Date Mar.	X Fineness	R Moving Range
0	1	2	3	0	1	2	3
1.	16	92.6	-	21.		92.7	0.5
2.	18	92.6	0.2	22.	26	90.4	2.3
3.		92.4	0.2	23.		94.2	3.8
4.		93.2	0.8	24.		93.6	0.6
5.	19	93.0	0.2	25.		91.8	1.8
6.		92.6	0.4	26.	27	92.8	1.0
7.		90.5	2.1	27.		93.0	0.2
8.		92.3	1.8	28.		94.0	1.0
9.	20	92.8	0.5	29.		93.3	0.7
10.	22	93.0	0.2	30.		92.6	0.7
11.	23	91.3	1.7	31.	28	92.80	0.2
12.		92.4	2.1	32.		92.80	0.0
13.		90.2	2.2	33.		92.60	0.2
14.		92.6	2.4	34.		92.50	0.1
15.	24	92.8	0.2	35.	29	92.60	0.1
16.		93.2	0.4	36.		94.00	0.4
17.		92.5	0.7	37.		91.70	1.3
18.		91.1	1.4	38.		93.40	1.7
19.	25	93.1	2.0	39.	30	92.60	0.8
20.		93.2	0.1	40.		92.80	0.2
				41.		93.60	0.8

Table T8.3 Pulverised Rock Fineness Passing Through 100 Mesh Screen.

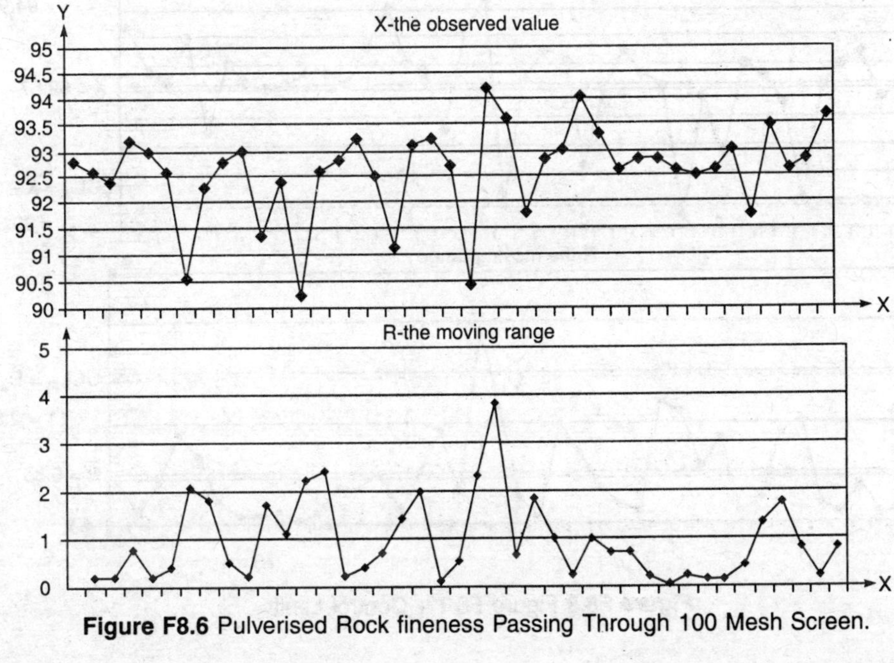

Figure F8.6 Pulverised Rock fineness Passing Through 100 Mesh Screen.

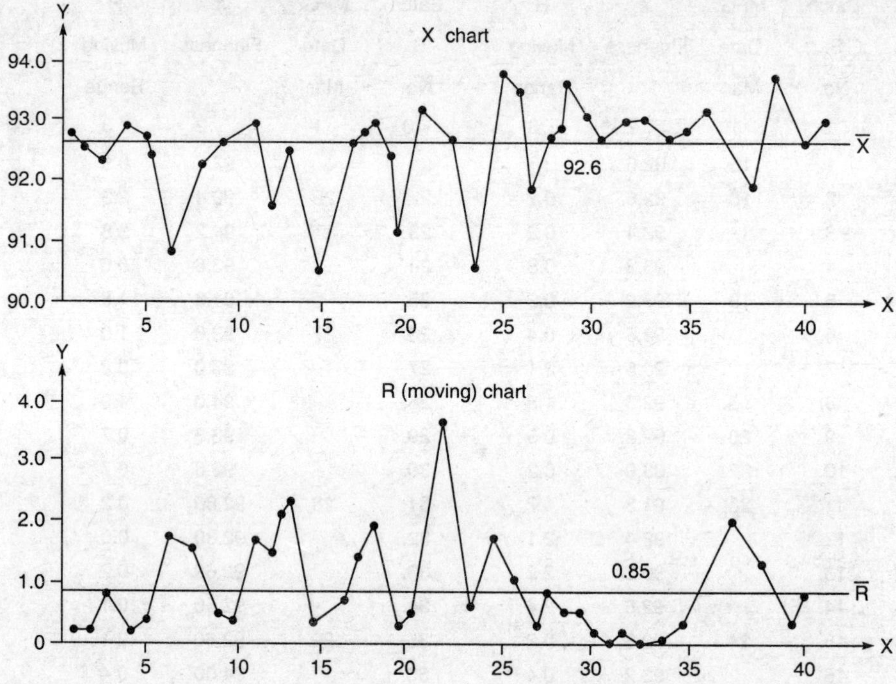

Figure F8.7 Pulverised rock fineness (passing through 100 mesh screen).

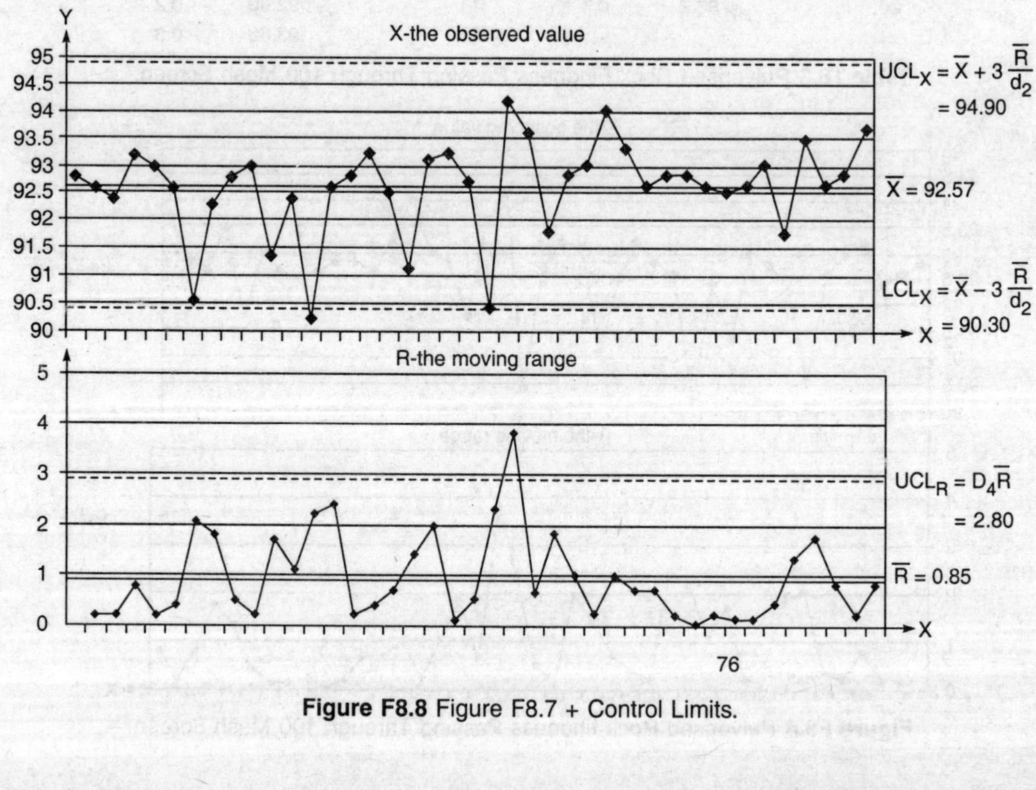

Figure F8.8 Figure F8.7 + Control Limits.

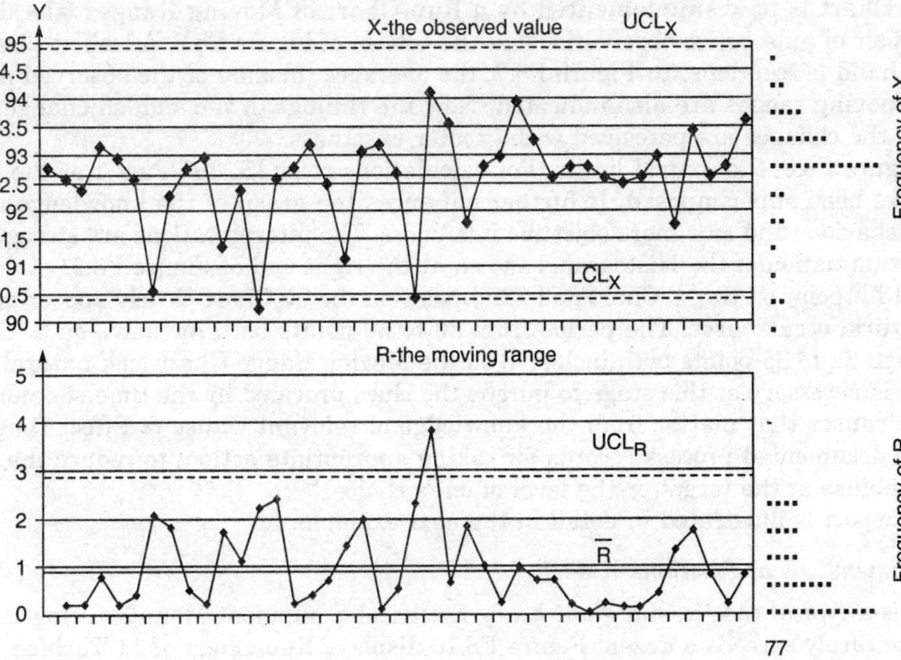

Figure F8.9 Figure F8.8 + Histograms.

APPENDIX: A8.1

$\Sigma x = 3702.8$, $n = 40$, $\bar{x} = 92.57 = 92.6$ Approx.

$\Sigma R = 37.0$, $\bar{R} = 0.925 = 0.92$ Approx., $D_4 = 3.267$, $UCL = D_4 \bar{R} = 3.04$, Note: $LCL = 0$
Range at serial no. 23 is 3.8. It exceeds the upper limit of 3.04. Ignoring this, the revised

$\bar{R} = 33.2 / 39 = 0.85$, $UCL = 2.8$ This time, none of the Ranges exceeds the upper limit. The ranges are thus homogeneous.

Process s = $\bar{R}/d_2 = 0.85 / 1.128 = 0.76$, $(d_2 = 1.128)$, $3s = 2.3$, $PC = 6s = 4.6$

$UCL\ x = \bar{x} + 3s = 92.6 + 2.3 = 94.9$, $LCL\ x = \bar{x} - 3s = 92.6 - 2.3 = 90.3$, $LSL = 90$

Desired $\bar{x} = 90 + 3s = 92.3$ for $C_{pk} = 1$ or 100%, Non-Conformance = 0.135 %
If $\bar{x} = 90 + 4s = 90 + 3.0 = 93$ when $C_{pk} = 1.33$, Non-Conformance = 32 ppm
Performance During *GOOD PERIOD is considered achievable and needs to be sustained.*

$\bar{R} = 2.4 / 8 = 0.3$, $D_4 \bar{R} = 0.3\ (3.267) = 1.0$, $s = \bar{R}/d_2 = 0.3 / 1.128 = 0.27$, $PC = 6s = 1.62$;

For $C_{pk} = 1.33$, $\bar{x} = 90 + 4s = 91$

Economy of $\bar{x} = 91$ versus 93 needs to be examined by comparing the extra cost and the likely rewards both tangible and intangible.

current $C_{pk} = (\bar{x} - LSL) / 3s = (92.5 - 90.0) / 2.3 = 1.1$ or 110%.
Assuming process to be in state of statistical control. It may be noted that it is not so.

N.B. During 28 to 30 Jan, fineness was unnecessarily recorded up to 2nd decimal place—of course the digit being zero always.

If $\bar{x} = 91$, $C_{pk} = (91.0 - 90.0) / 0.81 = 1.23$ or 123% (using the achievable value of $s = 0.27$)
$\bar{x} = 92.5$, $C_{pk} = (92.5 - 90.0) / 0.81 = 3.09$ or 309% (current average)
$\bar{x} = 93.0$, $C_{pk} = (93.0 - 90.0) / 0.81 = 3.70$ or 370% (average maintained occasionally)

This Chart is now supplemented by a Run-Chart of Moving Ranges (the difference between pair of consecutive observations) at the bottom of Figure F8.6. A look at this endorses our first hand perceptions. In Figure F8.7, the averages (means) of the observed parameter and the moving ranges are also indicated. Now the timings of the sudden changes and the degree of the changes are perceived with greater exactness.

In Figure F8.8, the control limits (For calculations refer IS: 397 Part 1 or the Appendix A8.1) have been superimposed. It further enhances the grasp of the knowledge about the process behaviour and aids more objective inferences. The interpretations are strengthened by critical examination of the Histograms shown at the right end of Figure F8.9.

The 13th point on the 'x' Chart and 22nd point on the 'R' Chart clearly appear as extreme or un-natural occurrences. The period from 30 to 36 points both inclusive on 'x' Chart and period from 28 to 35 points both inclusive on the Moving Range Chart look natural and very stable. It is necessary at this stage to pursue the clues provided by the time of occurrences to trace the causes that matter from the knowledge of relevant Cause & Effect Diagram and traceable documented process records for taking appropriate action, to reduce the variation and to stabilise at the target or the level of one's choice.

This aspect is illustrated in detail in the next example.

(b) KR Dimensions at Positions R & S.

This is a *typical* case from a *giant heavy* Electrical Equipment Manufacturing Unit where batch size rarely exceeds a dozen. Figure F8.10 displays Run-Chart of 11 Turbine Blades in production sequence with dimensions at positions R & S designed to be identical to conform to specified tolerances of 22.61 + 0.10. It is seen that none of the observations exceeds the upper limit of 22.71 mm or falls short of the lower limit of 22.61 mm and hence the complacency that all is '0 K' if not fine.

It is not realised that, glass of milk specified to be on the table, is not considered satisfactory unless placed at the centre or at least at a safe distances from its edges, if the table happens to be large enough.

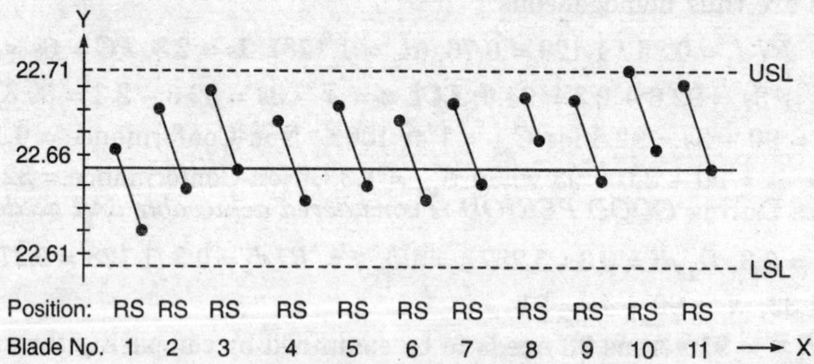

Figure F8.10 Run Chart.

The observation at position 'R' of 10th blade dot on upper specified value does not normally cause concern even though it is as serious a matter as the glass of milk at the edge of the table in real life situation, if not more.

A look at the Run-Chart Figure F8.10, reveals that:

* the observation at position 'R' is always more than that at position 'S', though both are designed to be identical.

* the differences between positions 'R' and 'S' vary from 0.03 mm to 0.05 mm. This makes up 50 percent of the permissible tolerance of 0.10 mm.
* the observations vary from 22.63 mm to 22.71 indicating a bit of higher process average than desired.

All these *symptoms* indicate presence of *systematic or non-random* sources of errors traceable to causes that are considered assignable or special or economically avoidable through appropriate remedial measures. This in fact reveals that potential for reducing error or variation in the process and hence improvement. A brain storming session among Cross Functional Team or group of members resulted in adequate work instructions shown at serial numbers 1 and 2 of column 1 of Table T8.4, *from the knowledge of associated causes and effects*. This may be considered as a simple version of Critical Failure Mode and Effect Analysis (CFM and EA).

Table T8.4 Work Instructions

Symptom	*Cause(s)*	*Remedy*
1	2	3
1. Dimension at position 'R' always higher than that at position 'S'.	1. Taper in the fixture 2. Incorrect cylindrical profile of the cutter.	1. Check taper with the feeler gauge and assure the use of correct fixture. 2. Check cylindrical profile and sharpen if required before use.
2. Incorrect level of setting.	1. Operator's strategy to play safe to avoid undersize/scrap.	1. Operator educated on capability and futility/loss of playing too safe as also the benefits of setting on target.
3. Large variation between blades.	1. Excessive run in roller.	1. Check roller run out and correct if needed.
4. Sudden shifts in dimension.	1. Presence of chips in the job.	1. Operator educated on the rewards of a little extra care to keep chips away versus prohibitive consequences.

All these measures were incorporated for the next job that followed. It may be noted that the specifications for the same operation for the new job for the same parameter are different; different target but same tolerance band. Tool or Fixture kit is not entirely the same. The results are shown in Figure F8.11. It is seen that the differences between the positions at 'R' and 'S' are almost eliminated, however the differences among the blades have increased, though yet not violating the specified boundaries. A repeat brain storming session culminated in updating of above Table T8.4 with additions at serial numbers 3 and 4.

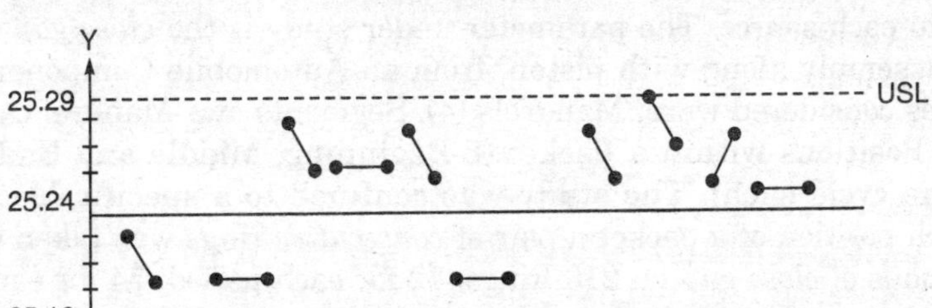

Figure F8.11 Run Chart.

The implementation of these measures for the next job resulted in blades with characteristic as shown in Figure F8.12.

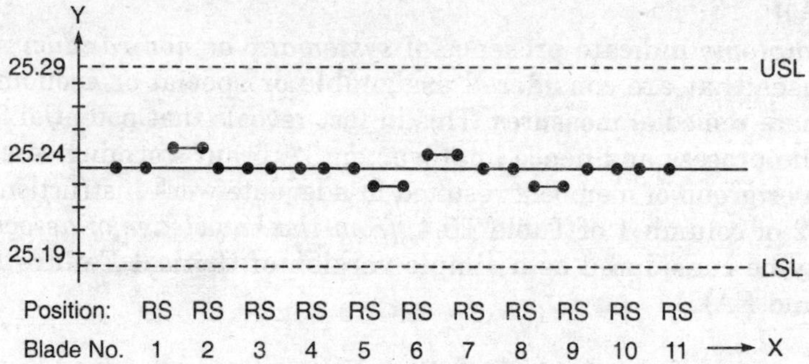

Fig. F8.12 Run Chart.

The Quality has tremendously improved. The average is very very (almost perfectly) close to the mean of permissible extremes. Also the spread around this mean is almost one-fourth if not one-fifth of that experienced hitherto. This implies five fold reduction in spread. Accepting concept of loss being proportional to the square of the deviation viz. Variance, the improvement is really worth 25 folds. Reduction in standard deviation to one-fifth is equivalent to reduction of variance to one twenty-fifth.

In fact variation has reduced to a level that the least count needs to be reduced and or the precision of measurement needs to be improved upon for effective control at the level achieved and initiating attempt for further improvement in future.

It is this sustained approach of pursuing clues to symptoms, symptoms to causes, and then on to sources of variation through Cause and Effect Diagram and associated Critical Failure Mode and Effect Analysis for perfection of specific guide lines to concerned human resource to facilitate production of product conforming to specified mean with least possible variation around it, that will result in alternative cycles of improvement and control. Once, these conditions are satisfied, the subsequent assembly operations will have none or least possible hazards.

(c) Automotive Components

This example differs from the previous two in the sense that data collection were pre-planned to examine the amount of variation contributed by various sources contemplated in the Cause & Effect Diagram. The balanced data facilitates, valid assessment of the contribution from each source. The parameter under study is the *close gap* of rings that form part of engine assembly along with piston, from an Automobile Component Manufacturing Unit. The sources considered were; Mandrels (4), Segments of a Mandrel called packets (3 on each mandrel), Positions within a Packet (3-Beginning, Middle and End) and Periods (3-consisting of one cycle each). The study was confined to a specific Model, Machine and Operator. At each position of a packet a pair of *consecutive rings* was taken as a sample. Thus we had observations of *close gap* on 216 Rings, 72 for each period, 54 for each Mandrel, 72 for each packet consisting of 24 for each its three positions.

The averages of these are respectively shown in Figure F8.13 with appropriate Control Limits. It is seen that:

- The three periods conform to the control limits, indicating steadiness of process over time. The three sigma control limits are approximately 0.015 Units apart. Since these are limits for averages of 72, the variation in individual observations are estimated as 0.015 ($\sqrt{72}$) = 0.1275 against Tolerance of 0.2, indicating a fairly high degree of process capability, an index of over 1.5. Such a real situation, is often responsible for complacency or de-motivation, thus forestalling any urge to even think for improvement leave aside any serious attempt in that direction.

* Overall average of Rings produced from Mandrel '1' possess less Gap than those produced from Mandrels 2, 3 and 4.
* The averages of positions of a packet show an increasing trend.
* The averages of packets of a Mandrel also show an increasing trend.
* The lowest average of about 0.415 corresponding to Packet A, position I almost assures that the gap of individual rings from among these shall fall short of specified minimum of 0.40. This could have been avoided with process mean of 0.50 instead of 0.457.

Process : Oval O.D. Turning Machine: M Operator: O Facilitator: F
Model: Ml Size: STD Parameter : Closed Gap Specification : 0.50 ± 0.10

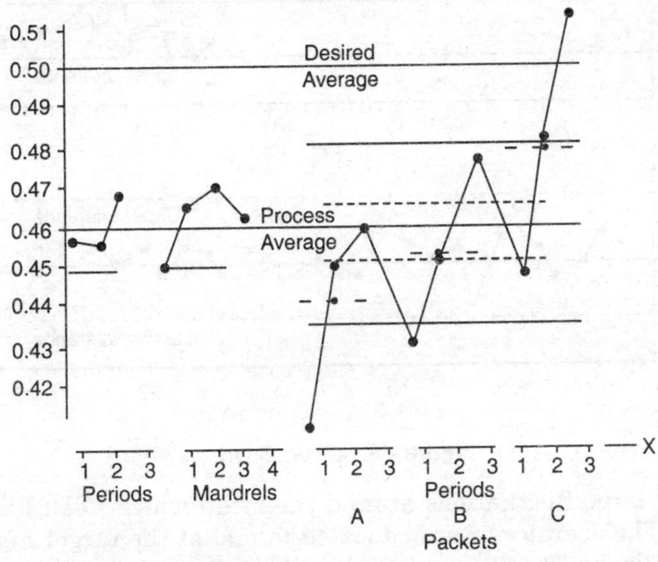

Figure F8.13 Oval O.D. Turning.

It is thus clearly seen that actions on:

* Proper maintenance of Mandrels (size and profile)
* Proper setting for the Gap size and
* Proper setting of Cutter Angle.

together shall reduce the Biases arising from differences among Mandrels and the trends between packets of a Mandrel as also among positions of a packet.

Thus the Quality gets improved both in terms of reduced variation and closeness to Target. Need, it be repeated that hitherto it was not considered necessary, since by and large the Gap rarely violated the specified tolerances and the knowledge about the contribution to variation from various sources was lacking. Graphical presentation of suitably collected data reveals both the *potential for improvement* and the possible approach to achieve the same.

Lastly a similar approach to monitor the 'OD' of Piston in the same manufacturing unit resulted in Improvement in three phases. See Figure F8.14.

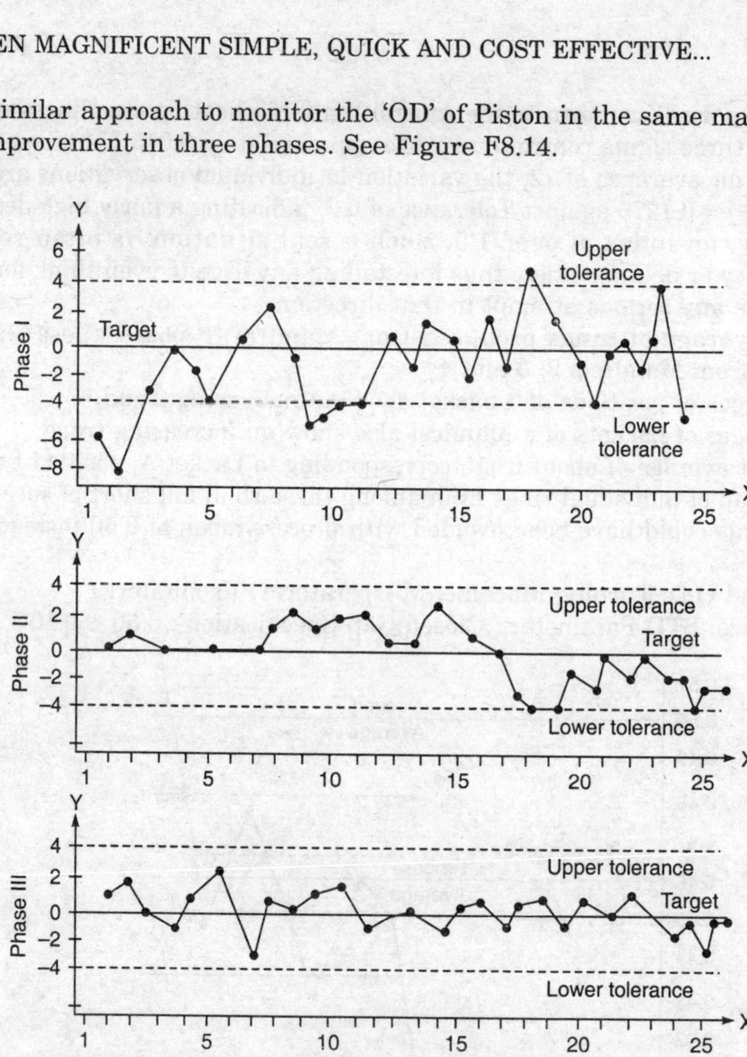

Figure F8.14 Run Chart.

Phase: I showed large fluctuations around the mean which itself has been frequently shifting over time. 'The operators were educated to aim at the target instead of anywhere within the tolerance limits. They were also advised not to reset the process in panic, though in good faith, but to wait for the signals from the limits, the trends and the associated unnatural patterns.

Phase: II shows improvement over Phase I, yet there are runs of large number of consecutive points above and below the specified mean and several points on the border line, though no violations beyond the tolerances. The lessons given at the end of Phase I were repeated and the desired cultural changes re-emphasised.

Phase: III shows excellent control, eliminating need for one hundred percent inspection.

Conclusion

It may be concluded from the above examples that a systematic approach to assess the total variation and its components from various likely sources paves the way for rational thinking of cost or effort to reduce variation from various alternative sources and alternative strategies versus gains in productivity of conforming products. By and large the extra costs, if any, are negligible in comparison to the returns.

(d) Heavy electrical—Rotor Disc

The shape of the rotor disc looks like the Figure F8.15. Only 4 to 6 units were manufactured in a year. This process was deemed non-repetitive and therefore considered not amenable to statistical process control. The job consists of drilling 48 holes. The job is loaded on the machine and appropriate settings made. Thereafter the same drill runs through all the segments to make a hole. This process is repeated 48 times to complete the job. To a statistician this constitutes enough repetition to permit use of techniques of statistical process control.

The problem is encountered in non-conformance of all of its parameters, such as diameter, pitch and concentricity of the holes. The data on all these parameters of all the holes (48 × 4 = 192) were asked for in the sequence of their operation, to examine the pattern of variation among (inter) and within (intra) segments. The management was serious about the problem and readily agreed to make available the data for the next job.

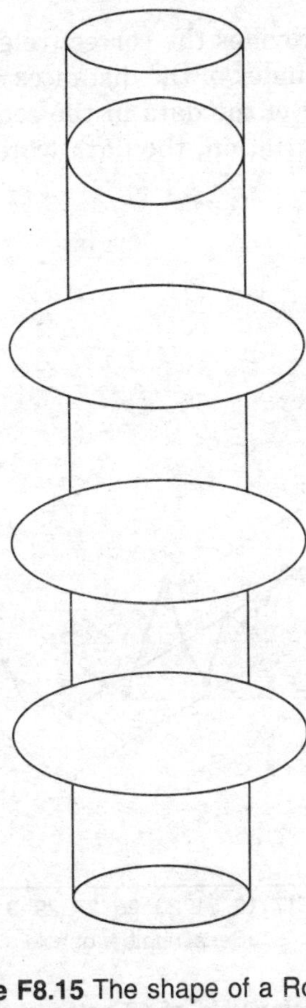

Figure F8.15 The shape of a Rotor Disc.

A lot many constraints and limitations did not permit data collection as envisaged and agreed upon. However data on *distance* 'd' of 48 holes from the periphery in production sequence was made available for the top segment, the cross section of which is shown in Figure F8.16.

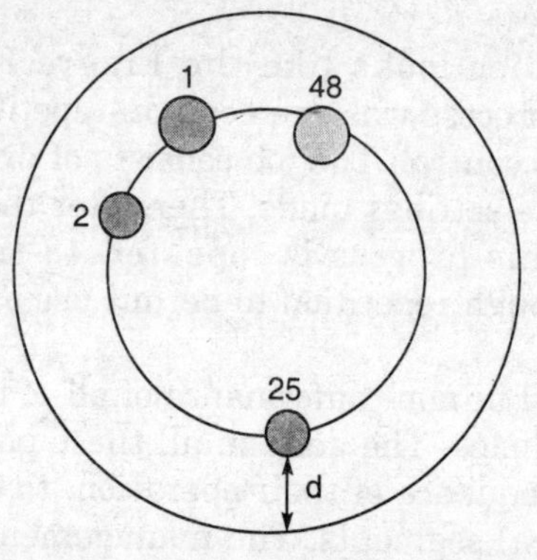

Figure F8.16 Cross section of top segment of a Rotor Disc showing location of 48 holes symmetrically placed.

Assuming that the periphery provides the correct reference with respect to the centre of the job, the *distance* 'd' provides an index of the distances of its centres from the centre of the job. With a pious hope that the plot of the data in the sequence of production might provide some vital clue to the pattern of variation, the data were plotted in a chart. The emerging diagram is shown in Figure F8.17.

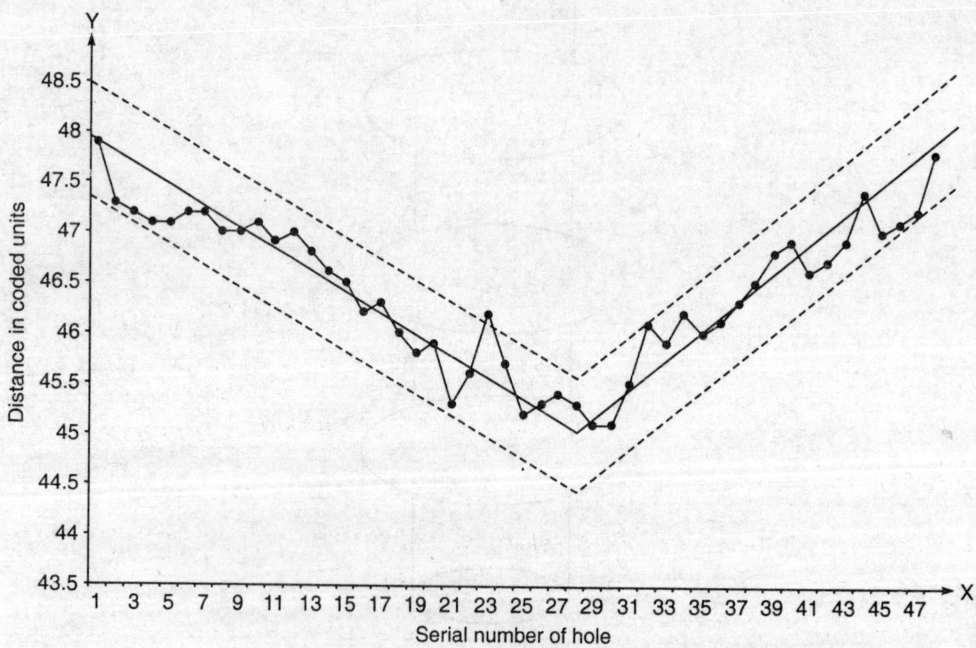

Figure F8.17 Run chart of distance 'd' of 48 holes of top segment of a rotor disc.

It is seen that:

* The difference between the largest value 47.9 and the smallest value 45.1 is 2.8 - approximately 3.
* If we try to draw a central line to the data, we find that the *distance* d has been decreasing in the first half and increasing in the second half.
* If we draw a pair of parallel lines, one on each side of the central lines forming a V shape, wide enough to just include all or most of the points, we find that the distance between the parallel lines is about 0.75. This is approximately one fourth of the observed maximum dispersion of 2.8 or about 3.

This leads to the conclusion that if one can trace the cause for this systematic trend and make correction for the same, the total spread or dispersion can be reduced to its one fourth or its variance to just one sixteenth.

The shop was visited to have a first hand knowledge of the process, to look for the cause of the systematic trends - decreasing followed by increasing. The process consisted of placing a standard template on the top of the job and setting the drill to make hole through all the segments, as positioned in the template.

It revealed that the template itself was not being placed appropriately, as per prescribed procedure. The placement was not properly centred. It resembled the situation shown in Figure F8.18. This explains beyond any shadow of doubt, the root cause of the typical pattern observed. It needs to be recognised that no other tool, how so ever complex, can provide clue to this kind of lapse. Run chart is one of the simplest, most inexpensive (all one needs is a paper and a pencil), and yet most powerful tool to expose the unwanted trends that provide clue to the deficiency in the process and hence paves the way to the solution.

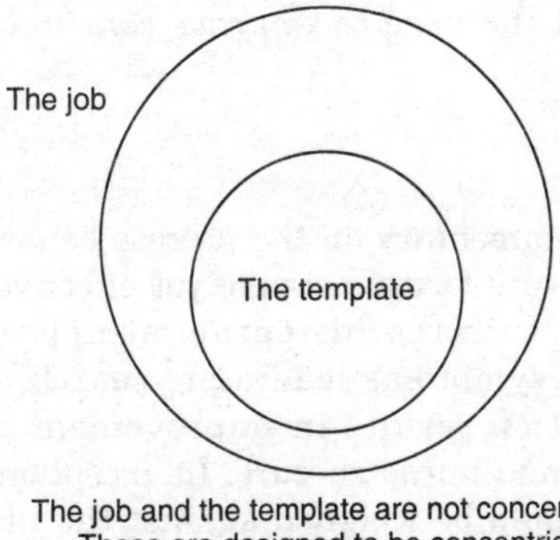

Figure F8.18 Comparison of desired versus actual fitment of template for drilling of holes.

Not only that, it was further seen that the template was designed to be located and fitted on the job with the help of two sets of nuts and bolts, placed diametrically opposite. The practice had been short cut and diluted to fastening only at one position on one side only, in good faith. The good intentions failed to recognise that in the game or the race of quality *short cuts are fatal.*

Once the correct regime was followed, the spread further reduced to one third. Thus, overall the spread could reduce to 1/12th or the variance by 1/144th. The tolerance was about one sixth of the observed spread providing an index of 0.16 of process performance. This could be enhanced to process capability index of about 2.0 making six sigma status a reality. This did not involve any extra investment.

Likewise there are several cases where the assessment of capability in a judicious manner and employing statistical process control compatible with its capability, in troubled areas have resulted in elimination of nonconformities (reduction from about 60 percent to zero percent) and one hundred percent inspection.

The axiom that *quality* demands voluntary dedicated commitment to understand the *right way* to do things and do *exactly the same,* to get the best possible from the available resources, gets reaffirmed. One needs to be passionate to give the best by *habit*. *Habits* die hard. Any improvements primarily coming from the new and supposedly better equipment is definitely short lived since we do not have the culture to sustain the way it ought to be run. We end up by adding to the costs without commensurate returns and increasing avoidable losses from lack of quality many folds.

Perhaps, because of this precisely, it is said that *quality is free*. *Passion* for *quality is spontaneous,* it cannot be purchased or procured against money. *In fact, Quality* is *cheaper* than *free* since no other investment gives such high returns. It ought to be remembered all the time that *doing things right does not cost even a paisa more while not doing right creates only losses, sometimes substantial and even fatal*. We are doing things anyway, all the time! Either *right or wrong*. *Wastage* of any resource is *sin*. The *most* precious and perishable resource is *time*. *Therefore, no idling* and *no wrong doings.* Of course the latter is worse than the former. Comparison is futile because both these habits are *fatal diseases.*

Run-Charts or suitable control charts as a means to monitor one's processes and use of *histograms* as a supplementary means for review with appropriate periodicity are simple, effective, efficient and indispensable tools at our service to proceed from *symptom* to *cause* and then on to *remedy* to fulfil the urge to *improve continually to fulfil the needs of the society*. There is no soft option.

8.6.3 Conclusion

Run chart provides a running commentary on the process behaviour just as ECG reflects one for the heart. There is no substitute for this simple yet effective tool for the study of process behaviour. Of course, coupled with the histogram at appropriate interval, it provides additional value-able clues for symptoms leading to search of remedial measures. Their confirmation and implementation results in improvement of quality, productivity and prosperity with same resources and infrastructure. In exceptional cases the pros and cons of investment required, if any, should be weighed against the likely returns, for taking right decision.

It has been experienced that in many cases, the processes considered incapable were found to be capable or highly capable. The exercise of their control provide necessary confidence among the human resource and the product. This often led to replacement of one hundred percent inspection by nominal confirmatory sample checks.

In many cases maintenance of run charts on *process parameters* of concern at appropriate stages and compatible intervals ensures *products* of desired characteristics that alone shall guaranty its *fitness for use.* Such examples are common in foundries, textiles, vanaspati and hydrogenated oils, sugar, cement, plating, chemicals and what not. One needs to be innovative in deriving more and more from the usage of this procedure. The skill shall mature with experience. *True documentation and adherence to optimal standard work practices are necessary to achieve excellence.*

9

A COMPOSITE COMPREHENSIVE CASE STUDY

9.1 PROBLEM

The problem, faced by a heavy electrical manufacturing plant, required identification of factors causing increase in failure rate of turbo generator bars during high voltage flash test from 0.3 to 4.8 percent and thus to eliminate the failures or at least to restore the status quo. The cost of each bar in late 70's of the 20th century was about a lakh of rupees. The nonconforming bar is a total loss. It cannot be repaired or salvaged. At best it may sell as a scrap at nominal value. The incidence of failure or nonconformance, in this respect in the works of the collaborators, is almost zero. These facts prompted the company to treat this problem on priority. This decision is based on the *PARETO* principle of choosing the problem that hurts most.

9.2 APPROACH

As a first step, a meeting of all technical experts from production and associated services like Design (product), Process technology (methodology), Maintenance, Testing was organized. Just by way of information the flow diagram is shown in Figure F9.1.

The brain storming session led to listing of the following 16 likely factors or causes or sources effecting the performance of turbo generator bars:

- *(i)* Season (Month)—Environment (Temperature)
- *(ii)* Shift of taping of insulation (Degree of supervision influenced by the team of executives and supervisors (technical personnel present)
- *(iii* to *viii)* Taping parameters—Dimensions—Height, Width and Periphery. Each Before and After curing—that is once after taping and again after curing with the identity maintained.
- *(ix)* Time lag between taping and curing stages of the process
- *(x)* Shift of curing
- *(xi)* Different types of Moulds
- *(xii)* Calibration of moulds—Pre and post calibration effect
- *(xiii)* Delta tan delta (Δ tan δ)
- *(xiv)* Time to fail (from start of testing or switching on to the failure of the failed bar)

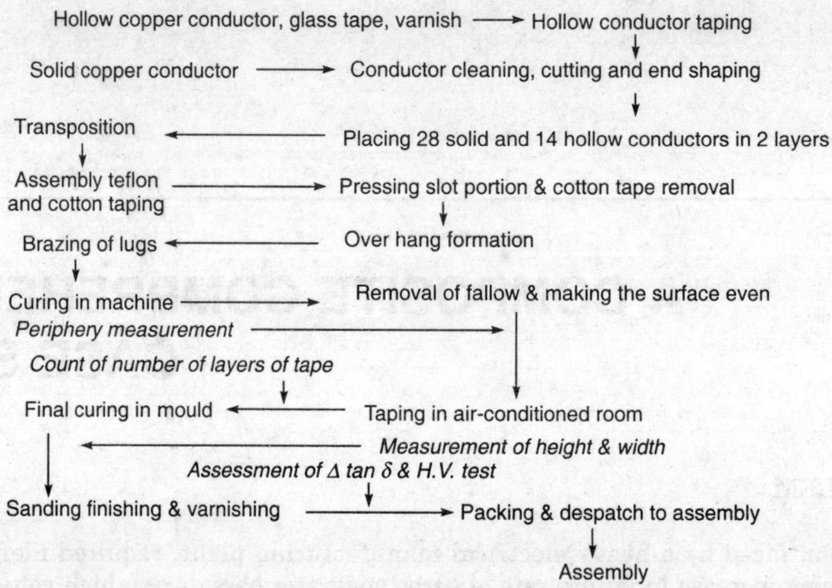

Figure F9.1 Flow diagram of the process of manufacture of TG Bars.

(xv) Location of failure

(xvi) Age of the insulation tape (the interval between the date of manufacture of the tape and its usage namely taping)

These are presented in the form of a *CAUSE AND EFFECT DIAGRAM* in Figure F9.2.

The process data had been recorded satisfactorily since inception, at respective stages. It was easy to transfer the same to one Master *CHECK SHEET,* shown in Table T9.1, developed for this purpose, for the desired period.

Even though, it is advisable to first look into the past data, as thoroughly as possible, to make a head start, yet as a short cut the use of techniques of Design of Experiments and Multiple Regression were contemplated. Inspite of the keenly interested involvement by the management, these two approaches could not be executed, for one reason or the other, such as constraints on existing facilities for testing and control, production rate and the production process. After loosing precious time, the team fell back on the first option, namely to examine the available past data.

9.3 DATA PLANNING

To make optimal use of the past data, it is desirable to plan its quantum and period.

To study the effect of season or month [cause *(i)*]: A minimum of two years data, month wise, are necessary. Latest two year data were compiled.

For causes *(ii)*, *(ix)* to *(xii)*, *(xv)* and *(xvi)*, which are of attribute kind, it was planned to use data for the latest one year only.

For the causes *(iii)* to *(viii)*, which are of measurable or variable type, it was planned to collect data for three months only; the best, the average and the worst months of the latest

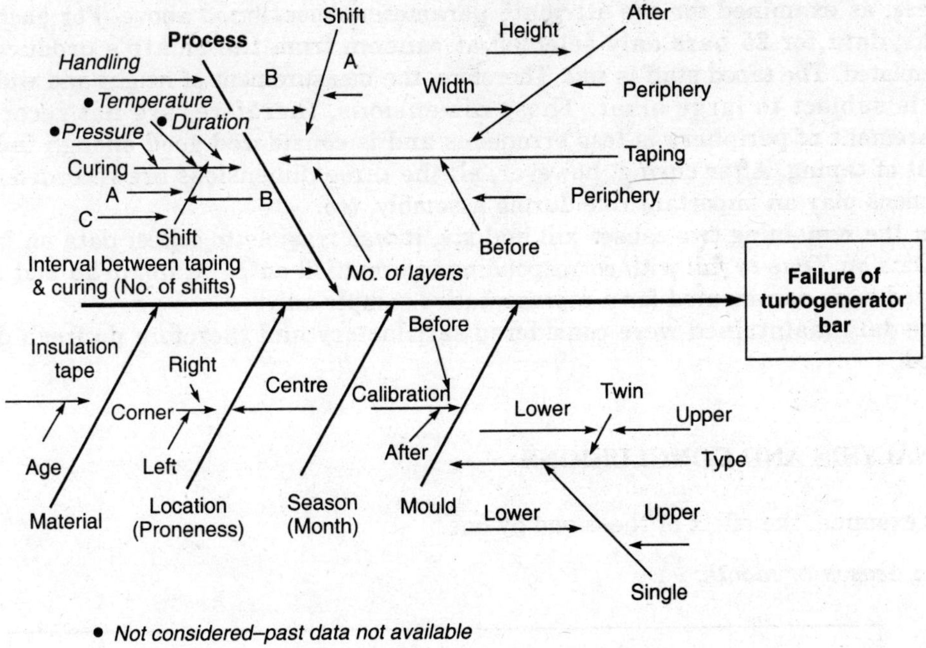

Figure F9.2 *CAUSE AND EFFECT DIAGRAM* for failure of turbo generator bar during High Voltage flash test.

Company: Department:

Bar number:
Date: Month: Year:

Shift of taping:
Date of manufacture of insulation tape:
Age of tape (days):
Before taping-periphery:
After taping-periphery:
Width:
Height: Supervisor

Shift of curing:
Interval between taping and curing [No. of shifts]
Mould type used: single/twin, lower/upper
Mould calibration date:
Supervisor

Testing: failure location:
Time to fail (sec.):
Voltage; Δ tan δ: Supervisor

Table T9.1 Master CHECK SHEET.

one year, as examined for the attribute parameters mentioned above. For each of these months, data for 25 bars only selected at random from the month's production were contemplated. The taped stuff is soft. Therefore, the measurement of height and width at this stage is subject to large error. These dimensions, therefore, are not recorded. The measurement of periphery is less erroneous and is considered good enough index of the amount of taping. After curing, however, all the three dimensions are recorded. All these dimensions play an important role during assembly, too.

For the remaining two causes xiii and xiv, it was feasible to collect data on failed bars only. Data on *Time to fail* with corresponding information on *Delta tan delta* and *Age of the tape* used were documented for a representative sample only.

The data maintained were considered satisfactory and therefore no fresh data were planned.

9.4 ANALYSIS AND CONCLUSIONS

Let us examine, the effect of these one by one.

Cause i, season or month:

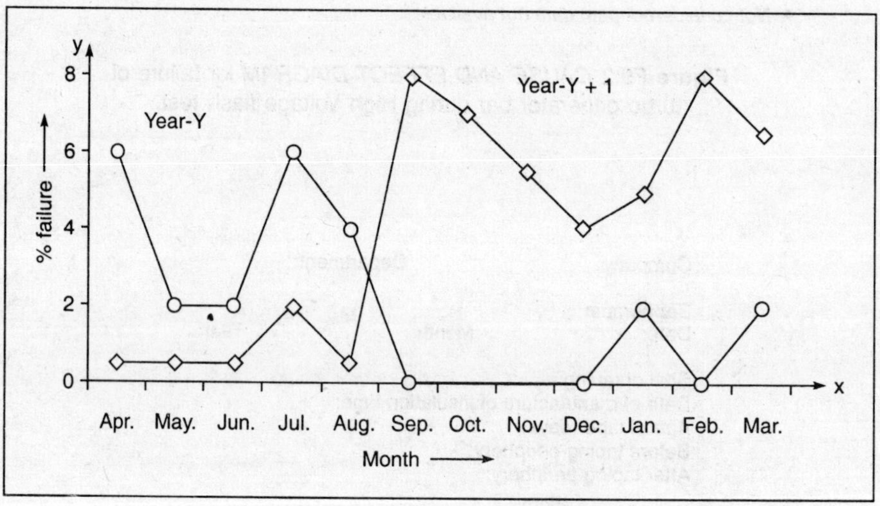

Figure F9.3 *RUN CHART* of failure rate of TG Bars for two consecutive years, month wise, to assess the effect of season.

Monthly *RUN CHART* of percent non conformance, in the form of line graph, for the two latest consecutive years, is shown in Figure F9.3. It is seen that in one year the period April to August is worse than the next seven months September to March. For the other year the picture is just the opposite, the first five months are relatively better and the next seven months worse. In case of any worthwhile dominance of the effect of the season or the month, the two annual graphs should be either congruent or at least parallel. In fact, if any thing, the reverse trend between the first five months and the next seven months, for the two years, contravenes the conjecture of any systematic effect of the months, beyond any shadow of doubt.

This finding, was not relished by the technical personnel, in the organization. They had literature from their collaborators stating that seasonal environments, temperature in particular does contribute to the incidence of failure during high voltage flash test. It took some time and effort to explain to them that the real implication of the finding is that, the contribution to the incidence of failure from some other factor(s) is relatively so high that the contribution of seasonal effect, if any, is relatively too small to be perceptible. In other words the effect of season, if any, is so dormant that it is camouflaged to such an extent by other more dominant factor(s), that it has become imperceptible. If the effect of dominant factor(s) is(are) taken care of by elimination or adequate reduction, then its effect might show up. The picture at the collaborators could well be different. The two situations are not exclusive to each other.

Cause ii, Shift of taping of insulation:

The taping of insulation is done in two shifts designated as A and B only, to the extent of production needs. The data on production and nonconforming number (failures), during the two shifts are presented in Table T9.2. This process is termed *STRATIFICATION* or classification by shifts.

Shift	Number		%	ldl	S	* d in units of s
	Produced	Non-conforming	Non-conforming			
1	2	3	4	5	6	7
A	735	26	3.54	0.20	0.66	0.30
B	224	6	2.68	0.66	1.20	0.55
Overall	959	32	3.34	0.00	0.45	0.00

P = 3.34, Q = 100-P = 96.66, PQ = 322.84, s = (PQ/n)
ldl = deviation s = standard deviation

* If d exeeds 2 but does not exeed 3, it is considered to have some mild influence.
If d exeeds 3, it is considered likely to have worthwhile influence.

Table T9.2 Nonconforming T G Bars *STRATIFIED* by shifts of taping.

It is seen that a total of 959 bars were produced during the year. 32, of these were found nonconforming. These failed during high voltage flash test. Thus, on an average, the nonconformance was 3.34 percent. Out of 959 bars, 735 had been processed during the shift A and the remaining 224 in shift B. The respective nonconforming units were 26 and 6 or 3.54 and 2.68 percent only. Now, let us take a bowl containing 1000 beads, 967 green and 33 (3.3%) red. Green represent good bars and the red the nonconforming bars. If, we pick up a sample of 735 beads at random, from the bowl, it is likely to contain 24 red beads in the long run on an average, if we repeat the process a large number of times. However, in any single trial, the number will vary marginally, this way or that way, due to what we may call a *sampling error.*

It will be nice to play this game or experiment to become familiar with or become aware of or gain first hand knowledge of the likely magnitude of *sampling errors,* which are very much a part of our day to day life. Soon, we shall realize, that the above deviations or differences for the situation on hand, namely samples of sizes 735 and 224 with respective nonconforming units of 26 and 6, are reasonably *small* and can, therefore jolly well occur as a normal phenomena and therefore can be given *benefit of doubt* as arising from sampling errors.

In fact, if we make appropriate calculations and refer statistical tables or ready reckoner for the purpose, the deviations of the magnitude, observed above, are likely to be exceeded on about 80 and 60 percent of the occasions respectively. For 95 percent confidence level, even differences up to twice this magnitude could be exempted. Therefore, there is no justification to look for the reasons for this difference between the shifts, as long its magnitude does not exceed the prevailing status. Any efforts to look for difference of this magnitude, are more likely to prove futile in our endeavour to find a solution to the problem on hand.. This way we can optimally use our human resource more effectively. We avoid spending our energy on futile exercises and conserve it for use on better occasions, when our mind is fresh to respond to the demands of the problem. On the contrary the matter is worth consideration if the difference exeeds the natural limit.

The executives did not consider this finding to their taste. Since, this process runs in two shifts namely, A and B only, when executives are present in varying number. They relented and suggested its examination over three shifts. The curing process extends to three shifts. Its examination is a part of this study. It is discussed under cause *x*, the shift of curing, subsequently.

Causes iii to viii—Taping Parameters—Periphery before curing, Height, Width and Periphery after curing:

As explained earlier, the latest year data were used to identify the three periods—the best, the average and the worst months. From the produce of each of these three months, a random sample of 25 bars was identified. The data on; the periphery, both before and after curing; and height & width only after curing was culled out for all the three periods. All the data thus obtained were summarized in twelve HISTOGRAMS. These are shown in Figure F9.4.

It is seen that:

* the histograms for height and width, after curing, for all the three months *worst, average and the best* are quite akin to each other.
* the histograms of the periphery after curing exhibit reduced spread in relation to those for the periphery before curing for all the three periods. This is expected to be so. The stuff is soft before curing and hard after curing. The measuring error is likely to be higher in the former case.
* the histograms of the periphery before curing show trend for successive reduced spread or variation over the three periods.
* the histogram of periphery after curing for the third period has distinct single peak in contrast to double peak for those for the previous two periods.

It may thus be concluded that if at all, *the periphery of taping may have some influence* on the quality of T G Bars with regard to its fitness to survive the high voltage flash test.

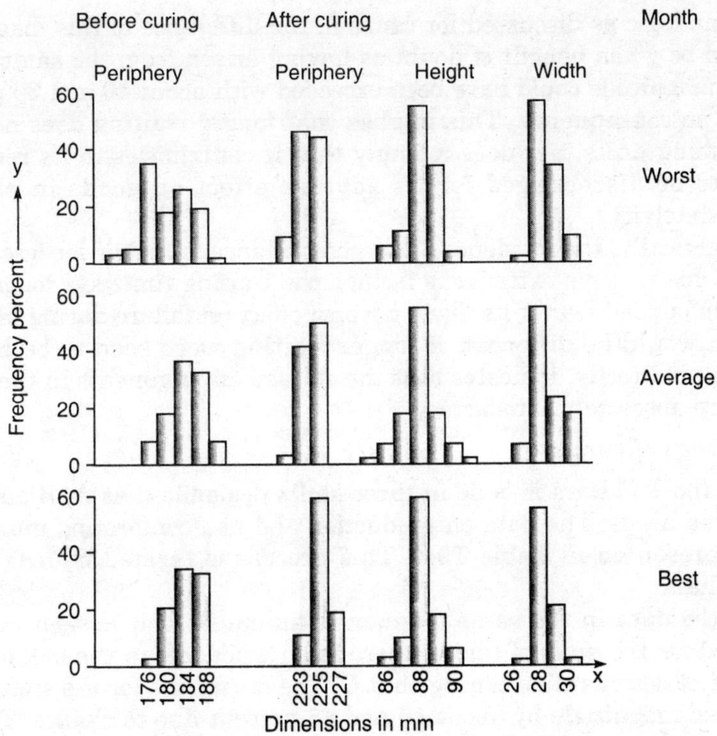

Figure F9.4 *HISTOGRAMS* of variable parameters of insulation of *Periphery before curing* and *Periphery, Height & Width after curing* for the 3 months, worst, average and best.

Time lag in shift	Number: Produced	Number: Non-conforming	% Non-conforming	|d|	s	d in units of s
1	2	3	4	5	6	7
< 2	546	22	4.03	0.69	0.77	0.90
= or > 2	413	10	2.42	0.92	0.88	1.05
Overall	959	32	3.34	0.00	0.45	0.00

Table T9.3 Nonconforming T G Bars STRATIFIED by Time lag between the processes of taping and curing.

Cause ix—the time lag between the processes of taping and curing

The data are suitably presented in Table T9.3.

This is *STRATIFICATION* by the duration of waiting. The dust is considered to be enemy number one of the electrical and electronic goods in process. The longer the waiting, the more are the chances for gathering the dust.

Using the same logic as discussed for cause ii, the difference of this magnitude between the two strata can be given benefit of doubt as having arisen from the sampling errors. The deviations of this magnitude could have been exceeded with about 40 and 30 percent chances. It is therefore of no consequence. This implies that longer waiting does not contribute to excess nonconforming units, nor does it imply that it contributes to its reduction. Longer waiting, needs to be discouraged for its adverse effect on goods in process and the consequential productivity.

However, numerically, the incidence of nonconformance is higher for lesser waiting. This assures us that if due to some extraneous factors, the waiting time gets longer, one need not have tension for unfounded fear of its likely adverse effect on failure during high voltage flash test. The concern would be different, if longer waiting were seen to be harmful. It also sufficiently, though indirectly, indicates that the storage environments in the air conditioned processing area are absolutely satisfactory.

Cause x—the shift of curing

The curing of the T G Bars is done in three shifts designated as A, B and C only, to the extent of production needs. The data on production and non-conforming number, during the three shifts are presented in Table T9.4. This process is termed *STRATIFICATION* or classification by shifts.

Interpreting the data in the same manner as for cause ii, it is seen in this case that, unlike as observed for the shift of taping there is an evidence, to the extent of about 98 % level of confidence, of deterioration during shift C. The deviations for the shifts A and B could exceed the observed magnitude by about 10 and 85 percent due to chance. The performance during shift C is almost twice as bad as the average of the three shifts. The hunch of the executives that when the cat is away the mice play or the *diluted supervision during the shift C may have an adverse effect on* the failure during high voltage flash test, is not completely unfounded. There might be some truth in it.

Shift	Number		%	|d|	s	d in units
	Produced	Non-conforming	Non-conforming			of s
1	2	3	4	5	6	7
A	459	9	1.96	1.38	0.84	1.64
B	255	8	3.14	0.20	1.13	0.18
C	245	15	6.12	2.78	1.15	2.42
Overall	959	32	3.34	0.00	0.45	0.00

Table T9.4 Nonconforming T G Bars *STRATIFIED by shifts of curing.*

Cause xi—the types of moulds

The nonconforming of T G Bars *STRATIFIED* by *the types of moulds* in use are shown in Table T9.5.

Using the same logic as in the interpretation of the impact of cause ii, it is seen that there is evidence, to the extent of 2% level of significance or 98% confidence, that *single upper*

Shift	Number		%	Idl	s	d in units of s
	Produced	Non-conforming	Non-conforming			
1	2	3	4	5	6	7
Single lower	370	8	2.16	1.18	0.93	1.27
Single upper	306	18	5.88	2.54	1.03	2.47
Twin lower	132	2	1.52	1.82	1.56	1.17
Twin upper	151	4	2.65	0.69	1.46	0.47
Overall	959	32	3.34	0.00	0.45	0.00

Note:
Mould single upper significant at 2% lavel.
It was checked since its cost is economical.

Table T9.5 Nonconforming T G Bars *STRATIFIED by mould type.*

mould might be contributing to higher rate of failures marginally. The deviations showing up for the rest of the moulds could have exceeded the observed values with the chances of about 20, 30 and 60 percent respectively. This Indicates that single upper mould might be a fit case for calibration.

Cause xii—impact of calibration of moulds

To study the presumable salutary impact of *calibration* on the failures observed during high voltage flash test, the calibration intervals were looked into. It was considered appropriate to *STRATIFY* the failures of T G Bars by their incidence during immediate *preceding* three months of the *calibration* date and *succeeding* three months. The data are accordingly presented in Table T9.6.

Using the same logic, as for, cause ii, it is inferred that the act of calibration did not do any good to the process. The performance was about the *same in both the situations,* namely

Calibration status	Number		%	Idl	s	d in units of s
	Produced	Non-conforming	Non-conforming			
1	2	3	4	5	6	7
Pre	214	9	4.21	0.51	1.45	0.35
Post	273	14	5.13	0.41	1.28	0.32
Overall	487	23	4.72	0.00	0.96	0.00

P = 4.72, Q = 100-P = 95.28, PQ = 449.722, $\sqrt{(PQ)}$ = 21.207, s = $\sqrt{(PQ/n)}$

Table T9.6 Nonconforming T G Bars STRATIFIED by pre and post calibration of the moulds.

before and after calibration. This implies that the decision to calibrate the mould was taken in panic, when one was not able to pin point a specific cause for increased incidence of failures during flash test. The action is taken on the basis of results during the test and this may not necessarily be linked to the current status in the curing area.

Causes xiii *and* xiv—*The impact of the value of* Δ *tan* δ *of the bar on the time elapsed till failure*

The data were available on the Δ *tan* δ *value* of the bar and its corresponding *time to fail in seconds* since switching on the current, for the failed bars only. It was decided to look into the relationship between the two, if any. For this purpose a *SCATTER DIAGRAM* was attempted, with the former on the x-axis and the latter on the y-axis. The outcome is shown in Figure F9.5.

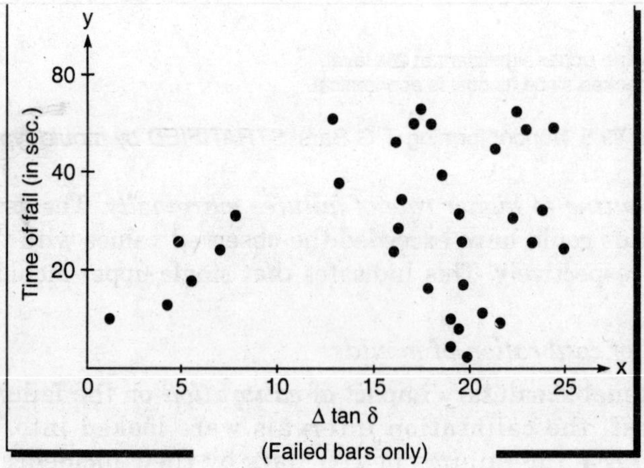

Figure F9.5 *SCATTER DIAGRAM of time to fail* versus its Δ *tan* δ value of the failed bars only.

It is un-fair, to look for the relationship for the truncated data, namely for the failed or the nonconforming bars, only. The constraints of the available data did not permit the look at the total picture. But with a view to salvage whatever information was possible, it was thought pertinent to examine whether the parameter *delta tan delta* influences even *the time to fail*.

The look as the Figure F9.5, shows that, by and large, the bars which have taken both less or more time to fail are also associated with both low and high values of delta tan delta. Thus in the light of above two considerations it is unfair to infer any association of Δ tan δ value of the bar, in its present range of variation, with its failure from the available data.

Cause xv—location of the failure

It was considered worthwhile to examine, whether any particular location(s) or segment(s) of the bar was(were) more prone to fail than the rest. If so, it might provide some clue(s) to work for the remedy. This art of looking at the concentration areas of the incidence of interest is called *geographical STRATIFICATION*. The outcome of this exercise is shown in Figure F9.6

It is seen that the portion along its flat length, which is taped mechanically, is free of any nonconformity. Out of 32 nonconforming units, nine are located near the left bend and the remaining twenty three are concentrated near the right bend. It may be mentioned that the taping near the bends is manual. Thus we may infer that there is a possibility that *manual taping near the bends* is prone to be more *weak,* than mechanical taping in the flat portion, resulting in the *failure* of the bars and *more* so near the *right bend* than the *left bend.*

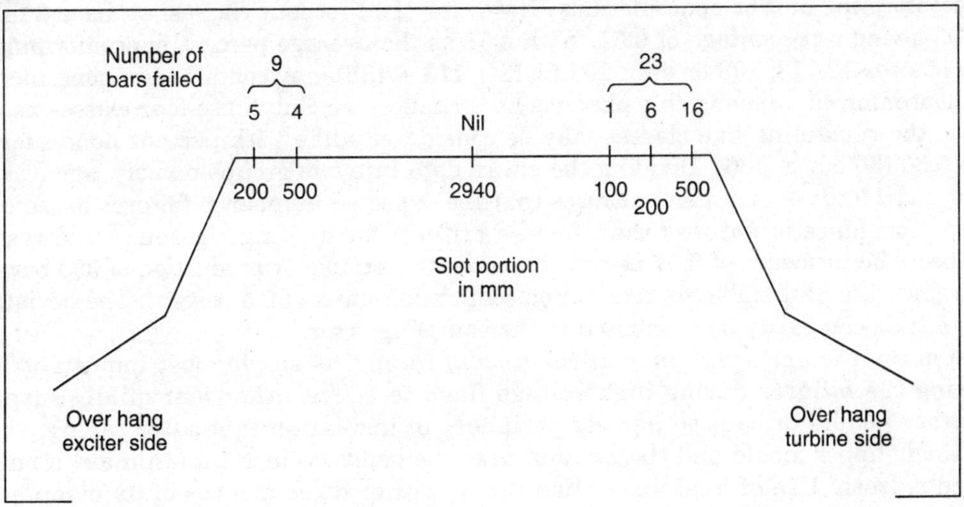

Figure F9.6 *Geographical STRATIFICATION,* showing concentration of location for the incidence of failures, along the profile of the bar, rendering it as nonconforming.

Cause xvi—Age of insulation tape

Lastly, the failures STRATIFIED by the age the insulation tape split into nine classes is shown in Table T9.7.

Age (in days)	Number Produced		Non-conforming		% Non-conforming		$\lvert d \rvert$		s		d in units of s	
1	2		3		4		5		6		7	
<50	12		0		0		3.34		5.19		0.64	
50-59	113	393	0	2	0	0.51	3.34	2.83	1.69	0.91	1.98	3.11
70-89	138		0		0		3.34		1.53		2.18	
90-109	130		2		1.54		1.80		1.58		1.14	
110-129	126		6		4.76		1.42		1.60		0.89	
130-149	97		5		5.15		1.81		1.82		0.99	
170-189	141	566	7	30	4.96	5.30	1.62	1.96	1.51	0.76	1.07	2.58
170-189	167		10		5.99		2.65		1.39		1.91	
= or > 190	35		2		5.71		2.37		3.04		0.78	
Overall	959		32		3.34		0.00		0.45		0.00	

Table T9.7 Nonconforming T G Bars *STRATIFIED* by the *age of the insulation tape* used.

It is seen that the failure rate is *zero or nil* as long as the insulation tape is less than ninety days old. It is 1.54 percent (2 out of 130) when the age is 90 to 109 days, both ends inclusive. It needs to be understood that if the lot contains zero nonconforming unit or non-conformity, the sample too shall contain none of these. On the contrary, if the sample contains zero nonconforming unit or nonconformity, it is not necessary that the lot too shall contain zero nonconforming unit or nonconformity. If we club the first four classes, we have 2 failures out of 393 giving a percentage of 0.51. With 0.51 as the average percent nonconforming, the samples of sizes 12,113, 138 or even 262 (= 12 + 113 + 138) may contain zero nonconforming unit or nonconformity. Giving this phenomena a chance, we club these four classes as alike. Like wise the remaining five classes may be considered alike with percent nonconforming units as 5.30 (30 out of 566). Dividing the entire data into two groups namely, age *less than 110 days* and *110 days* and *above* ensures that the expected number of failures based on the overall performance is not less than five—a criteria for making rigorous valid tests for significance. The incidence of 0.51 percent based on the sample or production of 393 bars does not contradict, the past stable percent of non conforming units of 0.3 percent. The deviation of this magnitude can safely be attributed to the sampling error.

To sum up, *the age of the tape at the time of taping,* is single most important factor influencing the failures during high voltage flash test. The other four mildly suspected contributory factors or causes; namely periphery of insulation tape after taping, shift of curing, single upper mould and the location near the bends; vanish into thin air, if only the tape used is fresh. Use of fresh insulation tapes, within three months of its manufacture, makes the process robust to take care of unintended imperfections arising from the four factors mentioned hitherto, inclusive of the skill of the workers doing manual taping in the curved zone of the T G Bars. To be on the safe side one may decide to *use the tape within three months of its manufacture.* The useful life at the collaborators end is one year.

9.6 THE SHELF LIFE OF THE INSULATION TAPE

The shelf life is a crucial factor. The tapes are required to be stored in air conditioned room and that too within specific temperature band. The storing area of the tapes was audited. It was found to be satisfactory. No flaw whatsoever was perceptible. The matter was probed further. The tapes are requisitioned from collaborators abroad. They intimate, the dispatch by air, as and when the supplies are made. The consignment is received at the airport by the team equipped with trucks full of bricks of ice, to be transported to the plant site about 200 kilometers away. There were occasions when the plane carrying the consignment is delayed and the availability of ice is not adequate. Even if sufficient ice were on hand, the vagaries of the environment may not permit sustaining the desired temperature in stipulated range of humidity. The problem thus reduces to improving conditions during transit of the insulated tapes to enhance its shelf life to one year or use it within three months or strike a compromise between these two extremes.

9.7 RECOMMENDATIONS AND CONFIRMATORY TRIALS

It was recommended to:

- use fresh tapes with shelf life less than 110 days.
- check single upper mould.

- maintain desired periphery during taping operation through adjustment or control of number of layers of the insulation tape.
- educate the workers on the *why* of conforming to product and process parameters or the tangible & intangible loss that their nonconformance can cause and the process knowledge of *how* to assure desired conformances. The workers need to be adequately facilitated and empowered.

As a long term strategy, consider measures to prolong the shelf life of the insulation tapes.

Immediately the confirmatory trials were started with the insulation tapes of shelf life up to 109 days. The older tapes were used for other purposes where requirements were not so stringent. During four months of this regime, 324 T G Bars had been produced. A total of four bars had failed the test including the two that got damaged in an accidental fall. Excluding these two, the failure rate works out to two out of 322 or a percent of 0.6. This is considered commensurate with the past best stable performance of 0.3 percent against the deteriorated level of 4.8 percent. Thus the old good status had been restored. The potential to achieve zero nonconforming is not a distant dream with another couple of iterations.

9.8 CONCLUSION

Thus the exploitation of the simple tools depends upon the ingenuity, skill and the technical knowledge of the user or the using team about the problem area. The advantages are boundless. It does require:

- element of *preplanning* of requisite relevant *data* with necessary *traceability,*
- collection of *adequate* (neither more than what is necessary for sufficient evidence nor less than that will keep the suspense on the conjectures or the hypothesis alive) data *economically,*
- data to be *true.*

This should be followed by prompt valid analysis or summary leading to recommendations that need to be confirmed through trials before implementation on routine bases. The gains accrued need to be sustained before attempting another breakthrough.

All these efforts and allied inputs are investments that pay rich dividends. The question is not whether we can afford these investments but whether we can do without these? The gains are made possible through commitment to quality or the needs of the customer. This is not possible through money alone. Of course, the invested seed money comes back many times over. The way to survive in this competitive world is putting one's heart and soul into one's mission qualitatively.

10

ORGANISING FOR QUALITY

9.1 NEED FOR QUALITY

Quality is something with which we are all concerned. Quality, of product, services and whatever we do, influences us all, in one way or the other, at all times and every where, from our health through activities to prosperity and even happiness.

Lack of quality impacts all of us. Tardy growth rate, rapidly growing unemployment and galloping inflation are some of its consequences. During the last three or four decades the exchange rate of Yen, the Japanese currency, has strengthened several folds in relation to US Dollar, while that of India has weakened several folds. The Yen and Rupee ratio has taken a steep dip.

Sloppy quality costs more through losses it inflicts. We suffer impact of poor communication among host of other essentials. The *inefficiency* of services is invariably attributed to *poor* quality of products or its *improper* use—both provide symptoms of *lack* of quality culture. Non-availability of funds is in turn offered as a convenient excuse.

Quality of product and services is indispensable for our survival.

10.2 REWARDS OF QUALITY

Against the background of losses caused by the lack of quality, the following have been experienced in Indian industries.

One cable manufacturing company estimated that in terms of economic viability, one percent reduction in nonconforming units of the product was equivalent to augmenting production by ten percent, which otherwise would require more investment in space, equipment and human resource. The alternative course of reduction in nonconformance through prevention of non-conformities offers a viable economic proposition. This motivated the human resource to bring nonconformance to one twentieth of the prevailing rate and thus ensure huge recurring savings at no extra cost.

Nonconformance of batches of disinfectant requiring reprocessing was reduced from thirty five to one percent in an insecticide production unit.

Rework of varying degree on all characteristics of a component was completely eliminated in an engineering organization facilitating assembly and enhancing product quality.

Nonconformities of over thirty percent during manufacture of blades of turbine in a plant manufacturing heavy electrical equipment was reduced to under one percent, in a span of about four weeks. This enhanced absolute production three folds and conforming production four folds.

Some state transport undertakings are providing better services at lower cost than some others.

Scrap generated during production in an electronic factory was reduced to one tenth of the prevailing rate.

Insulation property of ceramic insulators was improved to acceptable norms.

Admixture of glass particles in sealed bottles of a baby tonic were completely eliminated, assuring its safety for the humans.

The efforts needed for above achievements required little or no investment, yet yielded abundant returns, that could build up resources for further development at a faster rate.

Contrary to common belief that improving quality requires more costly inputs and reduced production, it instead begets enormous tangible and intangible rewards.

All these confirm that the national potential to achieve quality level of zero non conformance or parts per million or six sigma status exists. That too, at no extra cost or at economically viable cost. Can we afford to let go, this potential?

10.3 QUALITY IS THE BEST POLICY

From the foregoing we conclude that there is no substitute for quality. It pays in all situations, be it times of plenty or scarcity, inflationary or deflationary, protective, monopolistic or competitive market, since it always adds to profit by reducing manufacturing costs through prevention of nonconformities and avoidance of waste of resources in all its forms. The poor quality of work culture generates losses through enhanced rework, scrap, handling, production loss, customer dissatisfaction and associated intangible harm. The *salvation* from these ills *can be achieved by doing every thing right first time.* By preventing nonconformities we actualize *economic control of the quality of manufactured products and services provided. Thus Quality, is no doubt, an ALL WEATHER FRIEND.*

This leads us to the dictum, *take care of quality and quantity will take care of itself,* converse is not true. Hence as a policy, assign *top priority to quality.*

10.4 ESTIMATE OF POTENTIAL SAVINGS

An approximate idea of avoidable losses can be had quickly by the following methods of assessments.

Estimate the ideal minimal cost of product, likely to be incurred, assuming zero loss arising from wastage of material, under utilization of equipments and facilities including human resource. Compare this with the actual cost being incurred. The gap is the loss or the potential *diamond or even platinum* ready to be grabbed.

Estimate through simple work sampling study the ratio of activity time spent on doing fresh job to time spent on non fresh jobs in making the same worthy of acceptance. Divide the total cost of the infrastructure in the same ratio. The latter provides the tentative estimate being sought.

Compare consumption of each input like material, machine hour, man hour, capital etc per unit of conforming output over convenient periods for the recent past. Take the least of each of these as the achievable and work out the least possible, with these norms. The difference between this and the current cost gives us the estimate we are looking for.

Ultimately, one needs to organize data base that shall permit estimation of various elements of quality cost as per IS : 10708–1985 for review to plan and execute strategies for cost reduction. Fifty percent of the prevailing avoidable losses may be taken as an estimate of the immediate target of potential savings. It is invariably several folds of the profits being currently realized.

10.5 FUNCTIONS TO BE ADDRESSED

Attainment of quality cannot be left to chance. The cycle of improvement and maintenance of quality needs to be sustained through well planned and executed systematic activities which need to be updated through periodic review at appropriate intervals, depending on the situation on hand.

The activities may be divided among three functional groups—*PREVENTION, ASSURANCE AND IMPROVEMENT*.

PREVENTION through quality control measures:

Associated activities consist of

Assessing the needs of the customer and deploying these functionally.

Assessment of process capability and its optimal exploitation. Reference may be made to IS : 10645 and IS : 397 parts 0, 1, 2, 3 and 4.

Choosing the right suppliers.

Experimentation to develop fool proof product and process designs with minimal necessary effort and cost that assure valid reproducible results. Reference may be made to IS: 10427 parts 1, 2.

Planned collection of data and their appropriate analysis. Reference may be made to IS : 7200 parts 1, 2 and 3.

Training in statistical control methods for improvement and maintenance of quality of entire human resource in all segments and at all levels or as often said, both horizontally and vertically. Specially tailored syllabus is structured for each professional group. For example, the operators need to be educated particularly on product and process knowledge including remedial measures to cope with deviations that may arise, even though occasionally.

Inspection as a means for prevention, ASSURANCE and audit:

Inspection is not quality control. It provides important means for the same when its results are used to locate the sources of nonconformities to avoid their recurrence. It includes activities like:

Assessment, reduction, standardization and control of non-sampling inspection errors.

Selection of standard inspection plans to economically contain consumer's and producer's risks.

Inspection of incoming materials, in process and finished goods.

Inspection, during the process by the task performer himself, plays the role of prevention of the nonconformity and quality assurance. Such inspection takes three forms. These are *first off inspection, patrol inspection and last off inspection*.

- The *first off inspection* consists of inspecting the first unit or a small sample of consecutive units produced, after satisfactory completion of all preparatory work. This is done to confirm or ensure that the process *starts right*. Well begun is half done. In case of the unlikely event, that the first piece is not satisfactory, the process parameters are looked into to readjust them to the right levels before formal activation of the process.
- Having started right, the next step is to keep right. This is done through *patrol inspection* consisting of inspection at suitable regular intervals. The process is not disturbed as long as its produce is satisfactory. The recommended interval is half of the experienced stable running period of the process. As soon as the contrary event is encountered, the process is stopped. It is restarted only after satisfactory corrections have been made. Also the produce since the previous inspection is segregated to prevent a nonconforming unit reaching the customer. All the nonconforming units so found are acted upon in a manner to minimize the loss. This way, the patrol inspection helps us to continuously run the process rightly.
- Lastly, we inspect the last piece to ensure that the entire produce is satisfactory. Not only that, the set up is dismantled and all the jigs and fixtures used during the production are checked for their fitness for use before crediting to the stores. These activities constitute last off inspection. This avoids surprises often encountered when the production of this item is taken up next. Thus firefighting that hurt the system below the belt are avoided. These result in financial savings as tangible gains and peace as intangible bonus.

The above activities together constitute process control in action. This way alone, does inspection play the role of quality control. Or else it does only the post mortem. The inspection of a nonconforming unit is synonym with postmortem of the dead body. Often the loss or the harm that a nonconforming unit can cause is many times more. For example, a nonconforming unit of an automobile may cause an accident and many fatalities.

For salutary effect, the operator implying the entire human resource needs to be in *state of self control*. He needs to know the *why of the product* and *how of the process*. He should have *means to know* what his happening. He should have the *knowledge and means* to correct the process, in case an unwanted deviation appears. He should be adequately *facilitated* with necessary equipments and *empowered*.

* Maintenance and calibration of all inspection instruments, gauges, test equipments, tools, jigs fixtures etc is equally essential to accomplish the above functions as intended.

Assurance through Audit include activities such as:

Check inspection, Accuracy of inspectors, Analysis of customer complaints, survey of users experience, Assessing customer satisfaction and market quality determination.

Executive reports on quality with appropriate coverage, frequency and circulation

These activities ensure that the well thought out *quality plans are adequate and in place.*

The *planning* of all activities associated with the above functions may be centralized and *execution* decentralized.

For result oriented efforts, it is necessary to synchronise activities of all concerned. These should aim at achieving company goals. A *company manual on quality* serves a s a unifying force. It fulfills the following needs:

- Reference to policies and procedures with details of *what, when, where, who, whom, why and how*. This necessitates good advance thinking and assures accompanying benefits.

- Text for training.
- An aid to continuity of operations despite employees turnover.
- Basis for audit of current practices.
- Precedent for future decisions.

Lastly *IMPROVEMENT* is key to growth, the lack of which is fatal. All improvements are made possible through project by project studies. The group of workers accomplish the task with support from inter descipline teams and leadership from top management which ensures necessary facilities and empowerment.

10.6 INPUTS FOR QUALITY

Quality has been defined as *fitness for use.* This lays emphasis on *total satisfaction of the user* of the product or services. It encompasses:

Appearance : Good appearance makes *first sales.*

Functional : expected performance generates *repeat sales,*

Reliability : implying economic availability, maintainability and effective life cycle cost, *sustains the growth of market share.*

All these aspects are translated into specifications for material, process and product compatible with the required usage. Conformance to these standards in turn will assure fitness for use. It is an upward moving target. It is imperative that means to measure or quantify these are updated. These should be adequate, just good enough for the situation, neither too good that makes it more expensive and in-competitive in the market nor deficient that defeats the very purpose.

The input-output cycle is described in Figure F10.1. All inputs excepting men including women or human resource are inanimate. The wars have proved, repeatedly, if proof were needed, that it is the man behind the machine that matters. A well *educated, trained, skilled* and *committed HUMAN RESOURCE* of *high integrity* can assure appropriateness of all other inputs. Therefore the entire human resource besides being in *state of self control* in respective work area, ought to be adequately *facilitated* and *empowered.* Thus training and retraining assumes vital significance. One of the key factors for the success of the Japanese in industrial revolution has been *massive education* particularly in simple statistical methods in quality control on radio and T V network.

In India too use of mass media in education of human resource engaged in agriculture through programmes like *krishi darshan* has transformed the era of food *scarcity* to that of *plenty*. So much so, that other countries wanted the details of its curriculum. India rightly gave first *priority* to *food* front for *survival*. Now, we can look forward to *industrial revolution* for *prosperity*.

Success in this venture should be easier, since relatively industrial human resource is more literate. Even though massive efforts are involved on several fronts to reach the entire industrial human resource both horizontally and vertically in all segments, the mission of leading revolution to usher era of prosperity is reachable in foreseeable future.

10.7 TOTAL COMPANY WIDE ACTIVITY AND SELF QUALITY CONTROL

Activities need to be total and company wide that involve every one with mutual *unambiguous communication* and feed back based on strong *factual data.* Every individual should understand his role, do it religiously and keep it updated. This habit cannot be forced upon. It has to be voluntary, natural and self driven or motivated. This brings us to the concept of *self quality control*. It implies that, each one should understand one's role and perform it accordingly right first time. Of course, the human resource needs to be suitably *facilitated and empowered* for the success of the company wide mission of quality *leadership*.

Our *inconsistent attitudes* spell disaster for quality. One when we are responsible to produce goods or providing services and another when we are recipient of the same. These lead only to complaints and in house firefighting. A case, in example is, that millions of us working in *transport service* blame an equal number among us working in *telecommunication services* for its poor efficiency and vice-versa. Each group has many workable advices for the other. Remedy lies in each group organizing and practicing requisite measures in accomplishing the goals in respective work areas for the satisfaction of us all vitally concerned. How can one receive good product or service unless some one provides it? Charity begins at home, so does service start with self.

10.8 PRACTICE QUALITY FOR PERFECT QUALITY

It is well documented that there exist many situations in every plant which are not manifest but are causing avoidable colossal waste of resources. What needs to be recognized however is that there exists at least one *corner* in every plant, which is a symbol of excellence. Healthy practices of this corner need to be traced and emulated as a standard by all concerned, to successfully combat ill effects of the prevailing bad habits, to be worthy of leadership.

10.9 IS/ISO : 9000 FAMILY OF STANDARDS

It is a noble document. Its advice is simple, understand the needs of the customer including details of:

What does he want?
How much does he want?
Where does he want?
How much worth does he thinks it is? and
When does he want?
and go all out, intelligently to delight him viably.

The standard makes the use of statistical methods obligatory and mandatory.

It is experienced that managements going in for certification, find opportunities and or excuses to escape their use. On the contrary they should be looking for opportunities what else and where else can the simple statistical methods be used. The organization should imbibe statistical culture to own the quality culture. It should be using these tools because it helps them to perform better and better. The continued effort can transform them to be a leader. Then they are worthy of the certificate. Instead there is craze to manage to get the certificate as quickly as possible without necessarily developing the habit of improvements or mastering the culture expected of the organization by the standard.

Let the noble document intended as a guide be treated as such, with the reverence it deserves in one's own interest and the interest of the society at large.

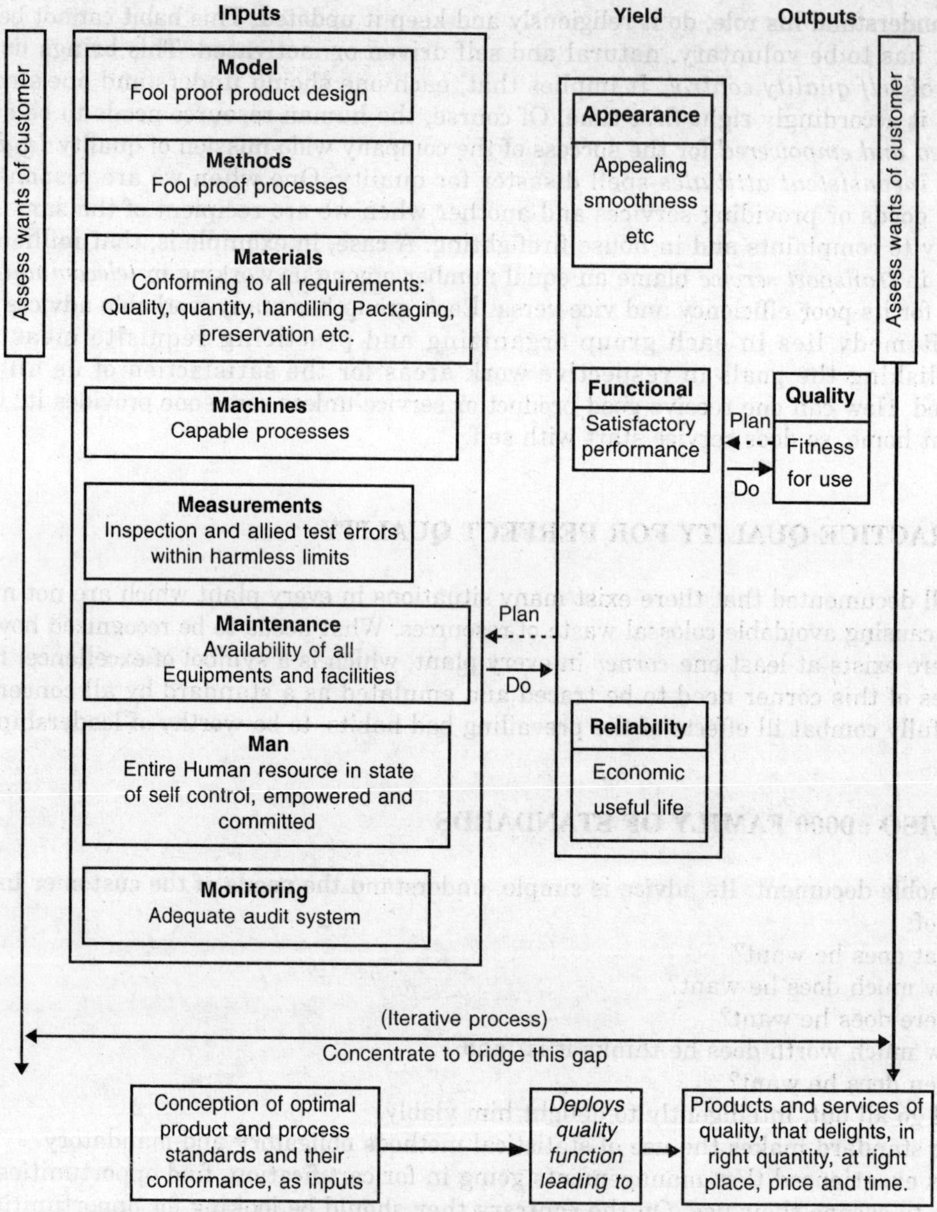

Figure F10.1 Functional Organization for Quality.

11

IMBIBING QUALITY AS A WAY OF LIFE THROUGH ITS ABC

An attempt has been made to enunciate basic tenets of quality and enumerate indispensable activities to put mission of quality culture for continuous improvement in place by using alphabets A to Z. Their accomplishment tends to attain leadership in one's field of activity.

A: ASSURANCE OF QUALITY IS TRULY POSSIBLE ONLY BY THE PRODUCER

Though the quality of sweets is enjoyed and judged by its takers, yet it is its maker alone, who knows its contents and the care taken during its preparation. Therefore, he alone is in a position to assure or guaranty the purity of its contents, its nutritive value, its harmlessness, shelf life and usefulness. Only the producer, the task performer or the server can build quality into the product manufactured or service rendered by him. He is knowledgeable about the degree of its fitness for use. He is the sole valid competent person to assure its real worth, no one else can. Since the knowledge of the former is primary and complete, while that of any body else is at best secondary and incomplete and so will be his assurance howsoever much well intended.

In early twenties, the link. between the producer and the consumer was direct or almost direct. The former understood the needs of the latter and took adequate care of it, for his own survival.

By mid twenties, a gap developed between the two, with the incoming of chain of production, inspection and marketing.

By late twenties, this gap widened further, by the augmentation of the chain by market research, design, process technology, packing, shipping and servicing teams.

Accordingly the concept of 'Assurance' by the supplier or the producer broadened to include every member of the team as the ultimate task performer. Each one needs to know his optimal role precisely and perform exactly the same and also remain updated. Each one's role is unassailable.

Any amount of audit or accredition cannot substitute the confidence that the primary task performer can exude. The audit is sample based. It is prone to both sampling and non-sampling errors. Their contributions can be estimated, reduced to harmless levels and controlled economically. These are rarely assessed. Often these are substantial. Their reduction, standardization and control go by default. It is more so, with respect to audits.

In this context, it will only be doing justice to ourselves and the society at large, if we get into the habit of doing self introspection for reforming ourselves rather than criticizing or finding fault with others for their correction. We need to fall in line with the dictum "charity begins at home", the law of nature, accepted universally. This is the proven dependable path for stable and steady continuous improvement leading to the, most sought after peace and prosperity.

ASSURANCE OF QUALITY IS ANTICIPATION AND PREVENTION,
OF ADVERSE EVENTS BY THE TASK PERFORMER'S ACTION.

B: BUSINESS STRATEGY OUGHT TO BE QUALITY FIRST

Quality is multidesciplined, multi pronged and multi edged sharpest weapon to minimize the costs and maximize the returns. It gallops profits for economic growth to combat inflation and unemployment, the twin chronic problems and thus ushers an era of prosperity. Quality is an all weather friend; be it era of scarcity or plenty, monopoly or competition, recession or inflation, war or peace, drought or floods; since it operates through optimal utilization of resources or minimizing if not avoiding waste in all its forms, thus reducing the costs.

Besides being all weather friend, quality is friend of 'one and all' of every segment of the society, be it marketing, designing, producing, servicing or consuming or whatever. In the long run, and that is what matters, what is not in the interest of one cannot be in the interest of the society.

Hence quality first, not quantity.
BEST IS NOT GOOD ENOUGH FOR OUR EXISTENCE,
CONTINUOUS IMPROVE-MENT NEEDS SUSTENANCE.

C: CUSTOMER FOCUS IS THE LIFE LINE OF QUALITY MISSION

Mahatma Gandhi had said:

A customer is the most important visitor on our premises.
He is not dependent on us. We are dependent on him.
He is not an interruption on our work. He is the purpose of it.
He is not an outsider on our business. He is a part of it.
We are not doing him a favour by serving him.
He is doing us a favour by giving us an opportunity to do so.

We often hear that the *customer is king or God. Do we ever revere or venerate him as such* or only dramatize for their make belief?

A *dissatisfied customer* rarely complains, dissuades many *others and himself* switches over. Every *single complaint compounds* to *over one* thousand alarms. *Is the complaint ever viewed as such* a monster?

No wonder, ISO: 9000 model for quality system starts with the assessment of the 'Needs' of the customer, urges its compliance and proceeds for evaluation of his satisfaction to focus on the gap to be bridged for its enhancement. All quality models worth their name need to do so with emphasis on continuous improvement.

The concept of customer has rightly developed from ultimate *user,* to include *next operator* in the chain and subsequently to *everyone else,* other than self. The self may be a customer of every one else.

The love for the *other* must be genuine. It should be selfless like that of a mother for her child. She anticipates the needs and leaves nothing to chance to fulfill these. She puts her heart and soul into her mission. She sacrifices every thing at her command, offers no excuses and makes no demands, yet achieves her mission of fulfilling even the implied needs of the child. Can such a love be a subject matter of assessment, audit or conformance?

The key is to make novel use of the resources available in fulfillment of the task or mission on hand, and not to cry about lack of any of these or offer lame excuses. One needs to remember and practice the dictum: 'A satisfied customer is the supplier's best unpaid salesman'. Can a certificate of conformance ever match the role of a satisfied customer?

CUSTOMER DESERVES SELFLESS LOVE, THAT OF A MOTHER, BESTOW IT.

D: DELIVERY IS A PERTINENT PARAMETER OF QUALITY

Delivery on time, is crucial to all types of operations. Each organization is both a supplier and a customer. Thus quality system is universally applicable. Initially the concept of just in time was focused on the delivery schedules only. A delayed delivery would upset production schedules, necessitating sudden changes on the shop floor. These add to the unproductive activities of human resource, which do not have any value addition. These also increase the idle time of equipments or machines. To counter this menace, an easy way of building inventory is resorted to. This also drastically adds to the costs. Together, these cripple the economy, some times even beyond salvation. *What use is the quality medicine that is sure to cure, if it is delivered after the patient is no more?* Optimization tools for scheduling inventory, production and maintenance, etc; are available to meet a variety of situations.

The concept of 'just in time' has been rightly extended to the execution of every activity, in the chain, as per optimal plan exactly - neither early, nor late, thus pegging the costs to be competitive. Just in time *leads to* or *is lead by* TQM, if only the attempt either way is whole hearted.

DELAYS IMPAIR BEYOND REDEMPTION, BETTER ENSURE ITS ABSTENTION.

E: ECONOMICS OF QUALITY IS A MYSTERIOUS MULTIPLIER

Lack of financial resources is voiced as the root cause for *POOR INDIA* failing to make a niche in the quality world, to justify the need for subsidy and commercial protections from the government for *SURVIVAL*. Host of other convenient, apparently logical and convincing arguments are advanced. Often these are only plausible excuses. Such measures only put survival at stake.

The best measure of economics of quality, easy to perceive, is the loss, both tangible and intangible, that the *lack of quality imparts* to the society. For example, a failure of a micro circuit breaker costing less than one hundred rupees may cause fire, resulting in loss of lives and property worth crores. A failure of a component of a locomotive may cause many fatalities and huge loss of equipment. Besides, delay in its restoration may jeoparadize availability of vital perishable goods in transit, with accompanying stakes for the life of the people. A failure of a social or business call may cause agony or loss of lifetime opportunity. A surgical error may cripple the patient and his family for life. A poorly washed linen may cause painful infection that may prove fatal or the treatment of which may cost fortune. Soldiers' life, war and freedom may be lost if shoe laces are weak. A failure in power generation or supply chain may bring the whole city or the entire nation to halt. Such examples of small errors or deficiencies in quality that make or mar are in galore.

Every business calls for some investment. The *returns* on investment in *quality* are *higher* than in any other form devoid of quality. These may vary from ten to forty times of the investment or even more. None of the countries having recognition in quality used subsidy as a stepping stone, nor the countries counting on subsidy have made a mark. Good quality always has a premium in the market; be it local, regional, national or international. Poor quality goes a begging sooner than later. A competent designer will give a better design at a lower cost to meet the specific needs of the customer and a competent processor will process it competitively on time to capture the market. Today, the world is shrinking and we are perforce a player in the global market.

Hence, it is a myth that quality costs more. In fact, its overall life cycle cost per unit of useful life is always lower. Experiences of successful organizations corroborate this reality. The rising inflation and unemployment and falling value of rupee are enough evidence of poor quality culture. The cost of poor quality, in the form of losses imparted, is colossal. Any poor country would have perished by now. Only India, endowed with the richest natural resources in the world by the *almighty,* could and has survived so far. Indians are among the poorest by dint of their *karma* or *complacency*. In contrast, though Japan is among the poorest in respect of its natural resources, the Japanese are among the richest by dint of change in their management style and work culture. India has the potential to be the world leader. Japanese quality and economy were in shambles in early twenties. If they could change, so can we and faster too. We must stop bragging about our plans and start working on these. The globalization phenomena would provide the needed impetus.

Statistically planned experiments lead to valid conclusions at reduced costs. These have been massively used by the developed countries to arrive at optimal, fool proof product and process designs *termed off line quality control*. These are strictly adhered to during manufacture *termed on line quality control* to deliver assured quality. These optimal levels be better adapted as product standards, *usually national* and process standards, *usually plant.*

The most frequently used designs are popularly called orthogonal arrays. The relevant theory was provided by Dr C R Rao, as early as 1947. He is the recipient of many National and International Awards and is popularly known as the best living statistician in the world. By introducing linear graphs and associated tables, Dr. G Taguchi has made the art of designing experiments, to meet the specific needs, easy. Its potential has yet to be realized in India. These methods provide both rapidity and economy. Experiments have made the journey, from stone age to space age, possible. Major era used classical methods of experiment. Statistically planned experiments have the potential to accelerate the *development* many folds.

EXPERIMENT FOR ADVANCEMENT.

F: FIXING THE PROBLEM AND NOT THE BLAME, IS THE NAME OF THE QUALITY GAME

Who created the non-conformity, is unimportant. How it happened is vital. His first hand knowledge as a spot witness is an asset for the working group popularly called *Quality Circles* or *Quality Improvement Teams or Zee Dee (zero defect) Committees* or six sigma groups in search of the remedy, through leads provided during their *brain storming sessions*.

Brain storming sessions provide wonderful opportunity to the team to work successfully to fix the problem and not dissipate energy in blaming, witch-hunting or fighting among selves. Anyone can commit an error. It is futile to trace to pin him down. There is need to unite to develop foolproof system and methodology and put these in place to forestall the recurrence of the problem. In this manner we march forward with conviction, smash every fresh problem encountered one by one and sustain growth rate of improvement. Statistics provides a kit of versatile cost effective tools to win the quality game.

FIX THE PROBLEM, NOT THE BLAME.

PARDON AND PARTICIPATION TO SOLVE IT, ARE ONLY NATURAL TO A SUCCESSFUL LEADER.

G: GOOD INTENTIONS, PIOUS ETHICS AND NOBLE MORALS CATALYZE SUCCESS

Just as healthy cultivation practices catalyse germination of seeds to maturity, in the presence of right doses of fertilizers, environments, pesticides, and water; *good intentions, pious ethics and noble deeds boost quality efforts in the presence of sound systems and apt methodology*. The latter are not sufficient by themselves. These do not substitute the intentions, deeds, honesty, and integrity of the human resource. However these do *influence the outcome* of his passion and efforts to the hilt.

Global competition, in the offing, will generate the will to kill (excel). Lack of it shall negate the growth to devastation.

GOAL OF QUALITY IS PRAGAMATIC, DYNAMIC AND RISING.

H: HUMAN RESOURCE IS VITAL ELEMENT OF QUALITY

All inputs of quality like model *design*, material, machine *equipment including cutters, fixtures and jigs, etc.,* method *process,* measuring devices *instruments and testing equipment,* maintenance *total,* marketing and management systems including monitoring are inanimate. In the event of an occurrence of a fault, the only animate *competent and committed human resource* can intervene timely to detect and correct the unwanted adverse change. This act prevents errors and constitutes genuine quality control. Thus indispensable human resource deserves parental development and sprucing up only. Gift it.

HUMANITY IS BEING HUMANE TO HUMANS.

RESPECTING HUMANITY IS RESPECTING SELF.

I: INSPECTION IS NOT QUALITY CONTROL. YET, IT IS ITS INDISPENSABLE TOOL

Inspection to detect non-conformities to identify conforming items and documentation of nature, severity and cause of nonconformity or its source is synonymous with *postmortem report* of the dead. It cannot remove the fault or revive the dead. A faulty item is worse than a corpse. It can cause many *fatalities*. Does this event attract that concern?

On the contrary, *self inspection* by the operator, the task performer, of each and every activity in his domain consisting of inputs, processes and outputs, to adhere to the prescribed optimal regime is the only sure way to *build quality* into one's performance. This type of inspection is *eternal vigilance* by the task performer. It *does not cost* even a single extra *paisa*. Yet, it is the *price* one has to pay. If there is any exception to this, it is not only worth it, but it shall prove the rule. A *committed vigilant worker* only can assure the *quality with pride*.

Direct cost of inspection consists of instrumentation, competent skilled human resource and allied facilities. Indirect cost includes that of inevitable associated activities like additional handling, storage, holdups, delays and rework as a part of *damage control* exercise. Together these can be colossal and even prohibitive. These make a *difference* between *success and failure*.

The results of inspection, form the basis of decision making. Therefore *assure the quality of inspection.* It should be compatible with the requirements of the situation, neither too good nor too bad and cost effective too. Optimally calibrated and maintained apt testing facilities with documented proven standard procedures, for use in the hands of competent human resource, are deemed assets to befit this need. Inspection errors, both sampling and non-sampling, like audit errors need to be *assessed, reduced to harmless levels and contained innovatively.* Statistical analysis of inspection data assess contributions from various sources to guide determination of strategic priorities for needful actions. Even simple decisions, whether to manage with sampling inspection or one hundred percent inspection or both, taken rationally are known to have contributed to substantial savings.

Culturally, the domestic kitchen is taken care of by housewives. They are not exposed to formal course in an academic institution on management of kitchen. They inherit the art of cooking and kitchen management by their ancestors. They provide the finest example of using self inspection as a tool to prevent a failure. They serve food to the family members to their taste without waste. They do not have a formal manual on system or procedures, yet they rightly and richly qualify for conformance to the requirements of I S O : 9000. They are accustomed to follow the right regime as a habit, so much so, that no conscious effort seems to be at work.

INSPECTION IS NECESSARY DEVIL, TAME IT, TO SLAVE IT.

J: JOY OF QUALITY IS ALL PERVASIVE

Joy of delivered quality is not limited to the customers alone, who are the main focus and beneficiary. Every constituent of every segments of the organization is joyful with pride for contributing directly or indirectly to the creation of such an item for the customer. They also share it with their families. Similarly, the *suppliers* and the *marketing agencies* including *retailers* without whom the *mission* would have been *incomplete* are party to this joy. Thus, the society as a whole is the beneficiary. It brings happiness and prosperity to all.

Monetary reward when shared by or distributed among more and more people gets reduced on percapita basis. Likewise the impact of sorrow also reduces when shared among friends, family members and will wishers. On the contrary the joy multiplies when shared.

JOY OF QUALITY IS UNIVERSAL. IT DELIGHTS ONE AND ALL.

K: KARMA, AND NOT THE SERMON, IS THE KEY FOR QUALITY PROGRESS

The Concise Oxford Dictionary defines *karma as sum of person's actions in one of his successive states of existence, viewed as deciding his fate in the next.* Further, the law of *karma* is the law of *cause and effect*. This law is scientific. It consists of two universal laws. First, *as we think so we become*. Second, *as we sow, so shall we reap*. All our thoughts, words, deeds, emotions, feelings and wishes are *seeds sown* in the *field of life*. In due course of time these seeds germinate, grow and bear fruits. Some of these are *sweet* and others *bitter*. All these to be eaten by us, none else. The effect of some causes is felt *immediately* and that of some others *later*. For example, the effect of over eating is felt soon after, while that of telling a lie may take some time. Remember *every cause has an effect and every effect a cause.* We need to be *judicious* in choosing causative factors and sustaining these at proper levels to get *desired effect* or result.

The elements of karma are the steps of the ladder to reach the rising goals of quality. Actions and deeds need to follow the piously thought out plans like a shadow. No gaps, because time is money and the delays can dodge the mission just as proverbially as *Many a slip between a cup and a lip. Time* is *instantly perishable* commodity. Catch it from the forelock. Pious thoughts that are documented, say, as a Policy and preserved for reference only to be produced on demand for special occasions like audits are not sufficient. It is just like having access to facilities treasured by the family for rituals on festive occasions only, to be preserved again and forgotten till the next occasion.

Quality is not an act of chance. It is the result of concious efforts to frame and execute a good plan. Next, these get done subconciously. To reach perfection, these should flow rightly, naturally and even unconciously.

KARMA, TRANSLATING WORDS INTO DEEDS, SPELLS MAGIC FOR SUCCESS.

L: LEADERSHIP FROM THE TOP

The top management must get passionately involved and lead the noble mission of excelling, so much so, that the rest feel their omnipresence. It should take no chance by leaving the job to the hired consultants, presumably though with good intentions. The role of consultants as trainers in theory and practice of methods is to catalyze. The company has to develop its own procedures and systems compatible with its culture and put these in place. This mission must involve the entire human resource or else it goes by default. This honourable task is neither delegate-able nor shareable.

Concurrently, the top management needs to avoid short cuts and short term policies. These are evidence of narrow mind and greed. The instant gains might benefit, at best, this generation only. On the contrary, there is need to depict a perspective long term vision to battle for ultimate victory for next generations, that is, our children, grand children and so on and on. Such should be the culture and tradition in place. Create a citadel that no one else can venture to encroach or destroy. Only parental care and support through examples and not precepts can make it happen. We have demonstrated our ability to plan and execute long term policies while discharging parental responsibility of developing our progeny at home. We need to extend this potential, beyond the four walls of our homes.

LONG TERM VISION IN SERVICE OF SOCIETY IS SURETY, THAT THE REWARDS BOOMERANG INTO PROSPERITY.

M: MOTIVATION FOR QUALITY

Motivation for quality emerges from awareness and knowledge of potential loss and harm that the lack of quality can cause to the society. By nature, one longs to project image of helpfulness. Any exception to this axiom, shall not lead one to commit sin of harming others. The philosophy; help ever, harm never is the key to happiness.

Thus, imparting knowledge of loss and harm likely to be caused by the deviation in any parameter of product, process or service in one's area of work is vital part of training to motivate. Any other form of motivation, particularly direct monetary incentives linked with performance is short lived. The salary is intended for performing to one's best ability. Can the lacunas be ever made up by extra reward or bribe?

Unexpected honour, recognition and promotions that follow quality performance, with additional responsibilities, accompanying status with allied benefits and above all, self contentment, though invisible, constitute long term gains and are most powerful motivators. This is reflected in the culture of the organizations. It is indexed by rare turnover of employees. People Join but do not quit, unless superannuated. Even their superannuation is delayed. These cannot be bound by rigid rules of law.

MOTIVATIONS THAT RECOGNIZE NOVEL CONTRIBUTIONS INNOVATIVELY WITH AFFECTIONATE GRACE, RESPECT AND HONOUR ONLY SUSTAIN.

N: NURTURE QUALITY CULTURE

Actions speak louder than words. Therefore, all actions must depict true commitment and respect for the enunciated quality policies, procedures and systems. Glamorous displays are short lived and suicidal. Development of quality culture is a mission to embrace the entire human resource with customer focus as the central theme. It requires immense patience and stamina. Remember, even hurry takes time. Yet, there is no room for complacency.

The *chalega* culture, the culture of accepting or tolerating the status as it is in vogue, has to go. Instead the culture of continuous improvement possible through chain of breakthrough studies should replace it. Globalization will put the challenge for survival in place. India has always proved its mettle whenever challenged.

Audits, as a rule or courtesy are carried out on prior intimation and planning. All of us remember our school audits or inspection days. The school is bedecked cosmetically as a bride to be and other preparations are made to put the best foot forward to allure the audit or inspection team for the day. The next day, the school is back to square one. It is also rare, that an eminent school with strong moral, academic and sports culture steals the show or gets recognized. Random audits do provide glimpses of the true culture in place. But these are forbidden. Bias does creep in. Unless every day can withstand the audit, without any special preparation for the day, whatsoever, the desired culture is not in place.

NEVER GIVE UP. MORE SO, THE MISSION OF IMPROVEMENT OF QUALITY.

O: OPERATOR IN 'STATE OF SELF CONTROL' ONLY, CAN DELIVER, GOODS AND SERVICES OF QUALITY

Everyone does work and is earnest to excel. If one performs the right way, it boosts the economy; and if one does the wrong way, it only plunders the economy. The problem boils down to knowing and practising the right method of doing things. Quality is an art of doing things better, in the sense of superior quality or lower costs or both, in relation to the competitor(s) at any given point of time or self over time.

This is possible only when every operator is in a *state of self control.* This is the only way to assure quality through prevention of non-conformity. The operator is deemed to be in the *state of self control* if he or she:

- is aware of the *why* of the quality needs. This implies that he or she has the requisite knowledge of the damage, harm and loss the lack of any desired quality feature can impart to the customer or the user in particular, and the society at large. In short, he or she knows the *why* of the product design.
- knows the *how* of the process, that is, what exactly needs to be done and the consequences of not adhering to the stated regime.
- has adequate means to perform the task assigned.
- has the knowledge and means to assess what is happening and its outcome, so that one can notice any adverse deviation as soon as it occurs.
- has the knowledge and means to correct the same.

Once the operator is in the state of self control as stated above, doing things right instead of wrong costs nothing extra. Yet, the former contributes to the profit and the latter to the loss.

For all practical purposes Company Wide Quality Control *CWQC,* Total Company Wide Quality Control *TCWQC,* Statistical Quality Control *SQC,* Total Quality Control *TQC,* Total Quality Management *TQM;* Total Integrated Quality Management *TIQM,* Synchronised and Integrated Total Quality Management *SITQM* or host of other similar nomenclature like JIT, Standardisation, Six sigma, SPC, MBO or VE imply the same mission. All these routes lead to the goal of prosperity provided these are practiced without leaving a gap and lip sympathy is given a go bye.

In this context, the term *operator* extends to include *every member* of the *human resource.* Every one needs to be trained, facilitated and empowered adequately to be able to perform right to accomplish the mission.

ORGANIZE TO FACILITATE AND EMPOWER THE HUMAN RESOURCE ADEQUATELY TO THE LEVEL OF SELF SUFFICIENCY, TO BE A WINNER.

P: PANACEA FOR SOCIOECONOMIC PROBLEMS LIES IN PRACTISING QUALITY AS A WAY OF LIFE

Practice of Quality, as a Way of Life is anticipating the problem and fixing it. In simple words, it is an act of prevention. Experience gained during transition from stone to space age has affirmed *prevention is better than cure* and *stitch in time saves nine.* The latter is suggestive of timely maintenance and that the returns on this are likely to be ten fold.

The true potential in India is many times over. In the eyes of the professionals from the developed nations, our plans are globally the best, while their execution among the worst.

Because of this dichotomy between Plans and Actions, it should be surprising if problems do not persist. These have, for over 50 years. One single major index is devaluation of rupee by over 10 and 20 times in relation to dollar and yen respectively since independence.

PLAN OPTIMALLY, PERFORM ACCORDINGLY, REVIEW AND IMPROVE.

Q: QUESTION REPEATEDLY TO ACCESS THE ROOT CAUSE OF THE PROBLEM

Ask questions that prompt journey to the initial cause of the adverse effect that needs to be removed. Ask *Why* again and again, till the root cause of the problem is reached. Its removal will foolproof the process to yield *Zero Nonconformity.* Besides *Why,* there are six more friends ready to serve this noble cause. These are *What, When, Where, Who, Whom and How.* Avoid the unfavourable events through fool proofing and cultivate the favourable ones to sustain the apt levels of the causative factors to advantage.

QUALITY FIRST, QUANTITY WILL FOLLOW. THE CONVERSE IS NOT TRUE.

R: RIGHT FIRST TIME AND EVERY TIME

Doing right first time is the most economical way to build quality into the product or service. This eliminates inspection, which is unproductive, nay, counter productive. It consumes valuable Effort, Facilities and Time. By doing right first time, associated hold- ups, rework and delays simply vanish.

The total savings on these counts, both direct and indirect augment power to compete and boost economy for substantial growth. The right way is unique. It is the optimal in the given situation. Each one in state of self control is its prerequisite, for one to be able to do right all the time.

RIGHT IS ALWAYS THE RIGHT CHOICE.
GIVE IT THE FIRST PREFERENCE.

S: STATISTICAL CULTURE IS THE BACKBONE OF QUALITY CULTURE

Statistics is a science in search of truth. It serves all other sciences. It is master of none. It delves into the void to know the unknown, to traverse from uncertainty to certainty. It is the key technology. It consists of formulating the problem rightly, gathering relevant and adequate data, aptly analyzing, validly concluding, making confirmatory trials and implementing these to reap the expected benefits and to repeat the cycle to continue the chain of improvements to reach the goal. Prof. P C Mahalonobis FRS founder director of the Indian Statistical Institute said so. He was a renowned Physicist, Economist, Architect of 2nd 5year Plan and Adviser to the Government of India.

Statistics provides means to estimate the deviations from the target, risks, errors or variations and the losses arising from these with desired confidence. Further analysis of these to gauge the contributions from various sources prioritises the approach to optimize. Statistics provides a kit of indispensable universal tools for feasible solutions to the problems. Beware, it's unscrupulous use casts shadow on its credibility.

ISO makes its use obligatory. Sorry, no option, better start wooing it to enjoy its beauty, support and enlightenment. Surely, it is dependable, fascinating, rewarding and trustworthy science.

STATISTICS IS SIMPLE, QUICK AND COST EFFECTIVE TECHNIC.
IT IS EASY TO LEARN, PRACTICE AND HARNESS TO CLICK.

T: TRAINING IS THE ESSENCE OF THE TOTAL QUALITY MISSION

Continuous training and retraining to update is the heart and soul of every quality activity. It is not confined to anyone discipline. To dominate in the war on poor quality, training, not only in job methodology and human relations, but also in intertwined disciplines like bench marking or goal setting, communication, cost analysis, decision making, environments, ergonomics, forecasting, health, house keeping, JIT; management by objectives, exception and flexibility; marketing; material management including storage, inventory and supplier's rating and development; motivation, operations research including optimization, PM, PERT & CPM, programming, queueing, reliability, scheduling, safety, standardization, systems; statistics including SPC, sampling and experimentation; terotechnology, time management, value analysis, waste control, and work sampling is essential. This list is not exhaustive. It ought to be multidisciplinary and cross functional. Each one needs to be proficient in one's domain. The field of training extends even to the compounds of the customers and the suppliers and constitutes an eternal loop.

Richness and accessibility of library facilities is a vital index of organization's love for growth and development of its human resource.

TOTAL INTEGRATED AND SYNCHRONISED MANAGEMENT FOR CONTINUOUS IMPROVEMENT, STARTS WITH TRAINING AND COHABITS WITH IT ETERNALLY.

U: ULTIMATE IN QUALITY IS EXCELLENCE

Excellent performance is possible only through pious thoughts and noble deeds. In other words quality is a way of life in service of mankind or society of which self, *each one of us,* is a part. One is a share holder in what one gives to the society. Do good, share its fruits. One cannot miss one's share in the outcome of errors made, whether intended or not. Therefore avoid errors. We are responsible for not only, what we give to others or society, but what we ourselves really are or what we receive. This is in tune with: *As we sow, so shall we reap.*

USHER AN ERA OF PROSPERITY AND HAPPINESS,
BY CRAVING FOR AND CARVING EXCELLENCE.

V: VALUE ENGINEERING IS THE PERFECT KEY TO QUALITY PROBLEMS

Value engineering provides very valuable guidance at every step by provoking analytical approach to choose from among the alternatives available, on the basis of net value added by considering cost versus benefit *including worth* ratio. Thus it provides right direction at every step. It renders timely help to avoid unintended pitfalls. Trust this friend. Do not spare him!

VALUE TRULY THE PROS AND CONS.
LEAVE NOT THE DECISIONS TO CHANCE.

W: WASTAGE SIGNALS QUALITY DEFICIENCY

Any resource that is improperly harnessed or under-utilised or idle contributes to the waste. Any amount of waste in any form, anywhcre at any time is unflinching evidence of deficiency in design of quality system or its operation or both. In contrast, *ZERO WASTE* is an evidence of an adequate quality system in place. It ensures prosperity through growth augmented by productivity, quality, value, worth and yield.

Wastes are generally invisible. These look more like a part of the necessary process and often go un-noticed. These offer precious opportunities for improvement. These opportunities get lost by default. An abrupt stock out in spite of excess store enhances production delays, work in progress and delivery schedule. Similarly, a sudden breakdown or a nonconformity observed at despatch station, causes enormous loss. Adverse chain of reaction sets in motion. These create unsurmountable loss of customer's goodwill and invite fire fighting. Yet all these are taken in as normal strides. Surprisingly, the total loss is large enough to win or lose. Its estimation is quick and easy. Calculate the ideal production, assuming every step was taken right first time. Compare it with the actual. The gap between the two is the waste. These are the gems in the waiting to be *culled.* Its *awareness* is *Quality Consciousness and pursuit Quality Control.*

Indians are among the best Fire Fighters. The joy of successful fire fighting blinds one to the loss from suspension of normal work in progress. It is not rare to see some one to create fire and extinguish it to earn promotion. Instead, the energy needs to be routed to plan and execute prevention of fire. Wise use of resources can reduce waste to zero. Unique use of all facilities at hand to succeed and not to find scape goat in the absence of any of these for the failure is quality control. Life is not possible without problem! Make it work for the solution. Consider two contrasting approaches to identical real life situations.

Two couples were blessed with a blind child each. One family, accepted the situation as the will of the God and showered well intended, all possible care on him. This pampering made him dependent on others. No activity, led to ill health. Doctors advised exercise. He was to run water hand pump, even though the family had servants to attend to such errands. The father in the other family happened to be an eye specialist. Hoping against hopes, he operated to the best of his ability. As expected, it did not succeed. The family assessed and exploited his other talents. He matured into a reputed singer. So much so he got an assignment in All India Radio. He gained job, delighted the society, gifted direct job to one person to help him in his daily chores, contributed to makers of musical instruments and transport services. He led normal life and did not invite mercy, sympathy or charity from others. Such incidents provide perspective of quality culture versus lack of it.

Let us recall other heroes. A blind worker in HMT Pinjore delivered higher production and superior quality among his colleagues. Kerala Cashew Industry engaged blind persons against reservation. To the surprise of one and all, they produced more and free of broken kernels too. Social bonds led them to marry among themselves. Blindness is not hereditary. They bore normal children and led normal family life, perhaps better. These heroes have demonstrated that they are second to none. They did not abuse the reservation privilege. They only needed equality of right to opportunity. They did not want charity. They fulfilled social obligations with flying colours by creating job opportunities for others, similarly handicapped.

In contrast the well intended reservation of 10 percent for 10 years introduced soon after independence has boomeranged. So much so that it has stretched up to 50 percent and in some states even up to 70 percent. Soon it may be one hundred percent! There is craze to access national assets and grab rights. The prime need is to, first fill the treasury by performing duties before exercising one's right to its resources. One is obliged to deposit before availing right to withdraw. The gear needs reversal. The father of the nation and other national leaders have been giving priority calls for doing our duties. We have yet to imbibe these.

Let us look at one more dimension of resource utilization. Everyone is gifted with large number of useful limbs. Do we make apt use of these to serve the society? Consider eyes, hands and legs for instance. Perhaps, it is a moot question, whether we use these to help the needy or to hurt the innocent. Once again these elucidate the mindset or attitude to quality as a way of life.

WASTE IS A CRIME AGAINST SOCIETY AND SIN AGAINST LAW OF DIVINE.

X: X-RAY OF ONGOING ACTIVITY EXPOSES POTENTIAL FOR IMPROVEMENT AND PROPOSES FIRST STEP TO ACCESS IT

Statistical process control (SPC) charts and allied analytical tools of shop floor data, synonymous with the *On Line Quality Control,* are akin to the terminology of X-Ray and ECG in the medical vocabulary. These provide true picture of current status of the health of the process and its owner. Thus, these aid timely detection and correction of adverse changes. Dr W A Shewhart pioneered SPC approach in his book *Economic Control of Quality of Manufactured Products.* Often, these lead to identification of sources which if acted upon aptly result in reduction of variation and hence, improved intrinsic process capability. Pursued steadfastly, the process capability may approach half of the specified tolerance range. The process capability index then equals two. This satisfies one of the two conditions of latest fad of six sigma status. The other condition is, that the process average of the controlled process, should not deviate from the target, by more than one-eighth of the tolerance, on either side. This, assures non-conformance of 3.4 ppm. If however, the average of the controlled process equals the target perfectly, the process assures non-conformance of only 0.1 ppm. One can determine the most economic target, if the tolerance is one sided or the losses from over size and under size are unequal.

This wonderful inexpensive proven tool has yet to be exploited to its full potential. The main reason for this lapse is that the relevant information on changes taking place in the likely causes of variation, documented in the cause and effect diagram, are not recorded as and when they occur or the recording is inadequate. Thus the vital data, the analysis of which could provide the solution, is missing. The culture of recording these vital facts needs to be imbibed. There is no time to lose.

X-RAY AND ECG REPORTS VIZ STATISTICAL PRESENTATION OF DATA, AID JOURNEY FROM SYMPTOMS THROUGH DIAGNOSIS TO CAUSE AND REMEDY.

Y: YIELD IS AN UNAMBIGUOUS INDEX OF QUALITY STATUS

Productivity, recovery or yield in the present context, implies *consumption* of any resource of concern *per unit of Good or Conforming* output. If the optimal capable system is in place, the waste of each resource will be minimal and the yield maximal. The cost of production shall be the lowest possible and this in turn shall make the enterpriser a leader in the market.

The ratios of the resources consumed to the conforming output will be smallest possible and their respective contributions to the production cost too shall be minimal. These ratios or their inverses are obviously apt indices of the quality status in place. The word *RESOURCE* has been used in broad sense to include tangible, intangible, direct and indirect inputs such as machine hour, man hour, energy and time. Time is invisible and instantly perishable input and hence the most precious. It deserves prime consideration continually.

YEARN FOR QUALITY, TO EARN PLENTY.

Z: ZERO DEVIATION FROM THE TARGET AND ZERO NONCONFORMITY CONSTITUTE THE IDEAL QUALITY THAT DELIGHTS THE CUSTOMER

It is not that *Zero Variation* and *Zero Nonconformity* are not possible. Where there is a will, there is a way. These are achievable and more often than not, these are achieved too. We fail to achieve when we compromise.

We never miss a flight or a train. We are never late. The consequences of being late guide our actions. We are late to our work places, because we compromise. Once again the consequences of being late to work weigh on our mind. This double standard is the bane for quality goals. Never the less, the achievements are ppm when it comes to missing a flight, collecting a lunch box from a customer's residence and delivering the same at his work place, preparing a salary bill or a meal.

However, the wonder of wonders is that, as soon as these are achieved, the expectations of the demanding customer rise instantly, leading to new challenges and targets, making quality improvement an eternal phenomenal game. Beware! There is no escape and no time to lose. With faith in the Almighty and confidence in self, we better start the journey at once to the best of our ability and integrity. Integrate and synchronize all the activities appropriately for desired results.

ZERO TO HERO THROUGH DEDICATION, ETHICS, INTEGRITY AND TECHNOLOGY.

It would be befitting to conclude with Swami Vivekananda's sermon:

DIVINITY GETS UNFOLDED INSIDE THE MIND OF HUMAN WHEN HE DOES IMPROVEMENT. HUMAN SOCIETY WILL BECOME DIVINE, IF ONLY ALL OF US CAN ACQUIRE THE HABIT OF DOING IMPROVEMENTS.

12

APTITUDE TEST

12.1 OBJECTIVE

The test has been compiled to enable the reader of the book and user of the tools to assess one's proficiency in understanding basic concepts of quality tenets and tools dealt with for making improvements through their application. This also enables one to identify the weak areas and the quantum gap that needs to be bridged. It may be desirable to time one self.

Each statement has four options. Read carefully and Circle, ◯ as the CAPITAL alphabet associated with the most appropriate alternative from among those listed. Correct, unattempted and incorrect responses invite scores of 4, 0 and -1 respectively.

12.2 TEST

1. Quality means,

A. Customer satisfaction.
B. Produce of proven product and process designs.
C. Product conforming to mid design specifications with least variation around it.
D. All the above.

2. Quality control means,

A. Prevention of nonconformity.
B. Detection and correction of adverse change.
C. Anticipating harmful effects of changes in inputs beyond control assessed through appropriate monitoring and manoeuvring the rest controllable to undo the same.
D. All the above.

3. Cause and Effect Diagram presents diagrammatically the causes, under appropriate heads and subheads which influence the effect of the problem under study. These causes are as,

A. arrived at a brain storming session among all those associated directly and indirectly.
B. listed in technology literature.
C. dictated by the technical chief.
D. short listed.

4. *Check sheets are necessary for,*

A. all the causes listed in the corresponding Cause & Effect Diagram.
B. all the causes impacting all the appropriate indices of the effect.
C. appropriate selected list from among A and B above. This may exclude such causes and effects which are definitely known to be constant.
D. none of the above.

5. *Pareto analysis is a technique to help identify the vital few,*

A. causes of nonconformities based on its frequency of occurrence.
B. causes of nonconformities based on the losses it imparts.
C. causes of nonconformities arising from negligence.
D. sources of zero nonconformity.

6. *Stratification is an art of collecting data such that,*

A. all details are available.
B. the guilty personnel can be identified,
C. the deficient inputs can be identified through grouping the data in a manner that inter group variation is large and intra group small.
D. all the above.

7. *Fundamental principles of stratification are,*

A. the corners of excellence in all places of work.
B. the availability of the data on the likely sources of deviations that are true and enable traceability.
C. consistent results that are repeatable under conditions of homogeneous inputs.
D. all the above.

8. *Scatter diagram is a tool to make a primary evaluation of the interdependence of two variables to assess,*

A. the model of interdependence.
B. the strength of interdependence.
C. likely potential gain from the control of the independent variable.
D. all the above.

9. *Scatter diagram needs to be drawn for two variables only when,*

A. relationship between the two is known to be linear.
B. nature and strength of relationship needs to be assessed.
C. two by two tabular or matrix data are seen.
D. none of the above.

10. *Histogram is a diagram that exhibits,*

A. the value on x-axis and its frequency on y-axis.
B. causes on x-axis and the loss on y-axis.
C. both A and B.
D. none of the above.

11. Histogram is likely to show,

A. central tendency and spread.
B. likely hood or otherwise of the capability of the process, given the specifications.
C. the extent of nonconforming units if the specifications are superimposed.
D. all the above.

12. Run chart shows,

A. the parameter of interest on y-axis and the sequence of production on x-axis.
B. the sequence of production on y-axis and the parameter of interest on x-axis.
C. both the above.
D. none of the above.

13. The prime responsibility for quality lies with the,

A. Top management personnel.
B. Primary workers.
C. Quality department personnel.
D. All the above.

14. Careful examination of run chart and histogram indicate periods influenced by,

A. chance causes only.
B. assignable causes only.
C. both A and B, to aid determination of favourable levels of the factors for attempting improvement.
D. none of the above.

15. For achieving good quality, the most important inputs are,

A. fool proof product and process designs.
B. competent, trained and committed human resource.
C. raw materials of good quality.
D. capable and well maintained machines including accessories.

16. Run chart as a tool is useful for,

A. only process control.
B. only process improvement.
C. both A and B.
D. none of the above.

17. For effective product control, it is necessary to,

A. assess the needs of the target customer segment.
B. develop fool proof product and process designs.
C. develop adequate systems for all associated activities and put these in place in the hands of competent empowered human resource.
D. perform all the above functions.

18. Better quality is always associated with,

A. high returns.
B. high cost.
C. low cost
D. none of the above.

19. Process control is achievable through,

A. First off inspection.
B. Patrol inspection.
C. Last off inspection.
D. all the above.

20. Pareto analysis is done through graphical presentation in the form of a bar chart and a cumulative percent curve with,

A. percent contribution on x-axis and causes on y-axis.
B. causes on x-axis and their contributions on y-axis.
C. causes arranged in non-ascending order on x-axis and their percent loss contributed on y-axis.
D. all the above.

21. To ensure hazard free subassembly and assembly, it is necessary to,

A. control the process of the capable machines to the desired target.
B. check all the components one hundred percent and use only those conforming to the tolerances.
C. use only the skilled operators, who are good at selective assembly and associated rectification jobs.
D. subcontract the job on piece rate basis.

22. Quality Management System and Methods aim at,

A. customer satisfaction.
B. conforming processes, products and or services.
C. conforming to legal and government regulations.
D. all the above.

23. Check sheets are planned formats to record observations made by,

A. visual examination, measurement, gauging or test.
B. visual examination only.
C. go-no-go pair of gauges and measurement only.
D. tests carried out on sophisticated equipment for critical parameters only.

24. Check sheets need to be designed to record data on,

A. critical parameters of vital products only.
B. any appropriate indices of inputs and outputs of concern of any activity of interest.
C. vital parameters of critical processes.
D. none of the above.

25. A good check sheet should,

A. be easy and speedy to record and summarise.
B. facilitate interpretation for planning and taking necessary action.
C. both A and B above.
D. none of the above.

26. Relationship between two variables x and y as seen from a scatter diagram is said to be positive, if

A. y increases as x decreases.
B. x increases as y decreases.
C. x increases as y increases.
D. none of the above.

27. Relationship between two variables x and y as seen from a scatter diagram is said to be negative, if

A. x and y increase simultaneously.
B. y increases as x decreases and vice versa.
C. x and y decrease simultaneously.
D. none of the above.

28. While drawing a scatter diagram, it is advisable to represent,

A. dependent variable on y-axis and independent on x-axis.
B. dependent variable on x-axis and independent on y-axis.
C. both A and B.
D. none of the above.

29. To improve visual appeal of a scatter diagram the scale for x and y variables should be such that,

A. the spread on x and y axes are as much equal as possible and fairly large for proper perception.
B. it is large for x and small for y.
C. it small for x and large for y.
D. none of the above.

30. If a histogram of 50 consecutive units produced with all controllable input parameters operating at constant level, shows a normal pattern, it may be inferred that the process is in state of,

A. economic control.
B. statistical control.
C. functional control.
D. objective control.

31. It is true that,

A. an histogram of any measurable parameter of any product shall always depict a normal pattern.
B. an histogram of any measurable process parameter shall depict a normal pattern.
C. if the process is in a state of statistical control and quality parameter variable (measurable) the resultant histogram shall reveal the normal pattern.
D. if an histogram shows a normal pattern, it is necessary that the data have emerged from a stable process.

32. If a scatter diagram shows no evidence of relationship between two variables y and x, while technologically a relationship is expected, it may be inferred that,

A. accuracy of data generated may not be sufficiently precise.
B. either x or y or both are controlled too rigidly.
C. y may be influenced more by some unsuspected factor other than the conjectured x.
D. any one or any combination of all the above.

33. Quality control means inspection of,

A. all parameters at all stages one hundred percent.
B. input and process parameters, preferably by the operator himself, to sustain these at appropriate levels, as means to control the process for controlling the product parameters for product assurance, at appropriate intervals
C. only critical parameters of critical stages of critical operations only.
D. every minute detail, very rigorously, by an independent agency.

34. For meaningful linear relationship, if n_1, n_2, n_3 and n_4 are the number of points in quadrants I, II, III and IV formed by the medians parallel to the axes x and y respectively, on a scatter diagram,

A. $n_1 = n_2 = n_3 = n_4$
B. $(n_1 + n_3)$ and $(n_2 + n_4)$ should differ by a wide margin.
C. $(n_1 + n_3) = (n_2 + n_4)$
D. $(n_1 + n_2)$ and $(n_3 + n_4)$ should differ by a wide margin.

35. The strength of a linear relationship as seen from a scatter diagram is judged by,

A. the slope of the line of best fit, the larger the better.
B. spread around the line of best fit, the smaller the better.
C. both A and B.
D. none of the above.

36. The best line of fit, in the context of linear relationship between two variables implies,

A. equal number of points on its either side.
B. least sum of squares of deviations between the observed values of y and its values as read from the line for the corresponding values of x.
C. least sum of deviations observed and predicted or estimated values.
D. none of the above.

37. The relationship determined from the scatter diagram between two variables x and y is,

A. limited to observed range of x.
B. limited to observed range of y.
C. both A and B.
D. valid universally for all values of x and y.

38. Any linear relationship needs to be fully exploited for economic gains by,

A. increasing or decreasing x according as the relation is positive or negative, indefinitely to extreme values.
B. confirming the nature of relationship beyond the observed ranges, till it shows up concave pit or convex peak for optimum exploitation.
C. increasing x and y simultaneously.
D. increasing x and decreasing y alternately.

39. If x and y are linearly related then,

A. increase in x will always be economical, if x and y are related positively.
B. decrease in x will always be economical, if x and y are related negatively.
C. increase in both x and y, if there is no evidence of any meaningful relationship.
D. economically viable decision will depend on the cost of increasing or decreasing x by a unit and the corresponding gain likely by the improvement in y depending on the slope of the regression line.

40. Given the specification of y of a product parameter, the specification of x for the process parameter showing linear relationship can be obtained by reading value of x, by projecting lines from the upper and lower specified values of y, parallel to x axis and reading the values of x corresponding to their points of intersection with,

A. upper and lower lines parallel to the regression lines, providing for process errors respectively.
B. regression line, the line of best fit.
C. lower and upper lines parallel to regression line providing for process errors respectively.
D. none of the above.

41. One of the key indices of the quality status of an organization is,

A. the growth rate of the market share.
B. the market share.
C. pending orders.
D. loan advanced.

42. Quality has been defined as fitness for use. It emphasizes that it is the customer who decides the degree to which fitness for use has been achieved. In this sense,

A. the customer is the next operator.
B. the customer is the wholesaler, the retailer and the end user.
C. the customers are the people around who are affected by it when the user is using the product.
D. all the above represent the customer.

43. Operator in the state of self control means that,

A. he knows the why of the product & what of the process he is performing and has the means to do the same.
B. he has the means to know what is happening as also the knowledge & means to correct when necessary and is empowered to do so.
C. both A and B above.
D. he has full authority to do anything in his area of work.

44. The cost of providing quality goods and services to the customer includes the costs of prevention, appraisal and failure. If prevention cost is increased to augment preventive efforts then the other costs are very likely to,

A. remain the same.
B. decrease.
C. increase.
D. increase or decrease such that the total cost will either remain the same or go up.

45. For the salutary effect of the application of the seven simple, quick and cost effective statistical tools, these must be used in the sequence in,

A. which these are published.
B. which these are discussed.
C. any combination and sequence best suited to the situation.
D. none of the above.

46. While conducting brain storming session for drawing cause and effect diagram, the leader should,

A. not interfere in a fashion that might discourage others from contributing or sharing their experience.
B. encourage and invite every one to contribute.
C. repeat exercise to arrive at total comprehensive picture.
D. take care of all the above.

47. The utility of the run chart is enhanced if the,

A. operator is penalized for the errors.
B. operator records all the changes occurring in the inputs and allied process parameters listed in the cause and effect diagram and the details of actions (nature and quantum) taken to remedy the adverse event in his domain of work.
C. operator checks and records only when he receives a complaint from the owner of the next process or any of the subsequent operations.
D. operator is paid on piece rate basis.

48. Quality improvement is synonym with,

A. continuous attempt to reduce deviation from the optimal target.
B. continuous thrust to approach zero nonconformity.
C. both A and B.
D. none of the above.

49. The operator wise stratification of nonconformities is attempted to,

A. identify the operators, responsible for the errors and losses, to hold them accountable for awarding appropriate punishment under the labour laws.
B. utilize their experience as first hand witness of the adverse situations that caused the errors, for making improvements.
C. benefit from both the above.
D. keep the operator under thumb, to discourage him from participating in union activities.

50. Interpretation of the process data with the aids of the run chart and corresponding histogram has potential to,

A. expose the scope for improvement.
B. propose first step in the right direction to reduce variation.
C. do both the above.
D. none of the above.

12.3 KEY TO THE ANSWERS TO THE OBJECTIVE TEST

1. D	11. D	21. A	31. C	41. A
2. D	12. A	22. A	32. D	42. D
3. A	13. A	23. A	33. B	43. C
4. C	14. C	24. B	34. B	44. B
5. B	15. B	25. C	35. B	45. C
6. C	16. C	26. C	36. B	46. D
7. D	17. D	27. B	37. C	47. B
8. D	18. A	28. A	38. B	48. C
9. B	19. D	29. A	39. D	49. B
10. A	20. C	30. B	40. A	50. C

LIST OF CASE STUDIES

Reference	*Industry Group*	*Product*	*Process*	*Item*	*Approach*	*Problem / Benefit*
5.51 a	Chemical	Vanaspati	Soldering	16.5 Kg Tin	Stratification by operator.	Elimination of soldering rework and reinspection.
b	Engineering	Tin manufacture	Inspection	16.5 Kg Tin	Stratification by inspector.	Elimination of customer return due to leakage.
c	Chemical	Unani pharmaceutical	Filling and labeling	Summer drink	Stratification by worker.	Out put doubled.
d	Electrical Engineering	Oil transformer	Assembly	Oil transformer	Stratification by group of workers.	Elimination of reassembly necessitated from leakage.
e	Export inspection	Service	Inspection	Many	Stratification by inspector and customer.	Judicious identification of pertinent complaints and inspections for rectification of situation.
5.52 a	Chemical	Rayon grade pulp	Chipping of Bamboos	Bamboo Chips	Stratification by Chipping machines.	Enhancement of production, improvement of quality and reduction in consumables.
b	Electrical Engineering	Capacitor	Filling and sealing	Capacitor	Stratification by pressing and sealing machines.	Reduction in rework to almost nil.
5.53 a	Textile	Cloth	Weaving Loom Shed	Cloth	Stratification by shifts.	Strategic planning to augment production and productivity.
b	Environmentally room temperature sensitive operations	Such as adhesive and sensitive auto parts	Heating etc.	Adhesive	Stratification by shifts.	Reduction in nonconformities and reprocessing.
c	Engineering	Large variety	Machining	Many	Stratification with respect to time to optimise tool grinding and change interval.	Elimination of non-conformity, rework and scrap.
d	Chemical	Electroplated parts	Electro-plating	Electroplated components	Stratification with respect to time to optimise interval for replenshing and changing plating bath solution.	Reduction in non-conformity and rework.

Reference	*Industry Group*	*Product*	*Process*	*Item*	*Approach*	*Problem/Benefit*
5.54 a	Pharmaceutical	Medicines Tonic etc.	Service/ marketing/ sales	Medicines	Geographical stratification for each product.	Enhancement of market share/sales.
b	Shoe manufacture	Shoes	Shoe making	Shoes	Geographical stratification of each type of non-conformity.	Increase of durability.
c	Engineering	Instruments	Assembly	Autometer watemeter watches	Geographical stratification for identify vital component and its parameter.	Reduced rework and scrap. Improved quality. Enhanced production.
d	Electrical	Fans	Assembly	Fan	Geographical stratification by location of testing/ inspection on ceiling.	Elimination of non-value addition rework due to notional slow speed.
e	Ware house	Service	Storage loss in transit	Food grains	Geographical stratification by despatching and receiving stations to identify routes prone to more transit loss.	Bias, among despatch and receipt stations identified, Improved conciousness resulting in reduced losses.
f	Hospital and Hotels	Service	Washing Laundry	Linen	Geographical stratification by source/work/usage area of linen.	Washing efficiency improved in terms of quality and cost.
g	Metallurgy	Castings, forgings, fabrications	Casting, forging, fabricating	Castings, forgings, fabrications	Geographical stratification by location for each non-conformity.	Reduced non-conformity.
h	Pharmaceuticals	Tablets	Tablet making	Tablet	Geographical stratification by level of mixture in hopper.	Hopper feed control interval standardised. Homogeneity of ingredients of tablet and its weight improved.

Reference	*Industry Group*	*Product*	*Process*	*Item*	*Approach*	*Problem / Benefit*
5.5 a	Chemical	Hydrogenated oil	Hydrogenation	Vanaspati	Stratification by supplier of oil.	Reprocessing eliminated.
b	Electrical	Fan	Painting	Blades canopy and motor body	Stratification by brand of paint.	Painting quality in terms of hardness and glossiness improved. Cost reduced.
c	Chemical	Rayon grade pulp	Chipping	Chips	Stratification by brands of V-belts for their useful life.	Chipping output increased and cost reduced.
d	Electrical	Fans	Service (marketing/ sales)	Fan	Stratification by dealers (customers).	Reduction in customer complaints and returns.
e	Pharmaceuticals	Baby tonic	Bottle washing, filling and sealing	Baby tonic	Stratification by process stage.	Elimination of glass specks, the presence of which can be fatal.
6.9 a	Jute mill	Jute woven	Weaving (looms)	Jute woven	Scatter diagram of output versus motor speed.	Optimal speed to maximise output.
b	Automobile	Vehicle	Running	Fuel (consumption)	Scatter diagram of fuel consumed versus running speed.	Optimal speed to-minimise fuel consumption and pollution and enhance longevity of vehicle and safety of passengers.
c	Chemical	Brewery	Brewing	Beer	Scatter diagram of output versus barley consumed.	Provided clue for need or opportunity for better process control or incompatibility of existing test procedures.
d	Engineering	Bicycle	Pedal and free wheel assembly	Non-free pedal movement	Scatter diagram of interference versus component parameter.	Design inadequacy identified.
e	Medical	Research	Patient care	(To assess) effect of drug	Scatter diagram of effect of each drug versus its dosage.	To decide on drug and its dosage for effective cure.

Reference	*Industry Group*	*Product*	*Process*	*Item*	*Approach*	*Problem/Benefit*
f	Steam boiler attached to hotel, hospital, vanaspati, textile chemicals etc.	Service	Boiling	Steam	Scatter diagrams of (i) Steam generated versus water consumed with due regard to temperature. (iii) Steam generated versus coal (fuel consumed) (iii) Output versus steam consumed.	Improved efficencies of steam generation and consumption.
g	Rubber manufacture	Shock absorbers for rail wagons	Curing	Rubberised plates/discs.	Scatter diagram of percent non-conformity versus amount of rubber compound.	Reduced rework.
h	Vanaspati	Hydrogenated oil	Bleaching	Hydrogenated oil	Scatter diagram of final color versus initial color of oil for imported and indigenous bleaching agent separately.	Reduced reprocessing and enhanced final production.
i	Heavy electrical engineering	Turbine	Curing	Rotol slot trough	Scatter diagram of product parameter versus input or process parameter.	Reduced scrap/rework.
j	Chemical	Titanium product and cement	Rotory kiln operation	Titanium dioxide	Scatter diagram of production/product parameter versus one by one parameter of feed and process.	Improved quality and enhanced production.

Reference	*Industry Group*	*Product*	*Process*	*Item*	*Approach*	*Problem / Benefit*
k	Rubber manufacture	Cycle tyre	Drawing or tread shaping	Tread	Scatter diagram of weight per unit length versus adjustment parameters like height.	End waste and process inspection reduced.
7.5 a	Engineering	Machine tools Aeronautics	Job shop	Many	Histogram for each dimension of component of concern.	Reduced scrap and assembly HICCUPS.
b	Chemical	Thermos flask	Blowing	Glass shell	Histogram of output (number produced) per shift.	Stream lining team work with potential to augment production by 30 percent per unit of time.
c	Chemical	Organic	Boiling	Organic compound	Histogram of production per hour over days.	
d	Engineering Electrical	Condenser plate	Drilling	Hole	Histogram of diameters of holes in a plate.	Reduced rework. Improved consistency. Enhanced production.
e	Plastic textile	Wound bobbins	Winding	Bobbins	Histogram of diameter of bobbins.	Improvement in validity of inspection and consistency of diameter of wound bobbins.
f	Engineering	Service	Inspection for amperes of a component	Component for Teleprinter	Histograms of amperes observed on same components by production and inspection personnel.	Inspection standardised to avoid wrong decisions on acceptance or rejection benefitting producer and consumer.
g	Engineering	Shaving product	Stamping and honing	Shaving blade	Histograms for centrality and honing angle.	Exploded myth of rejection due to old poor capable machines. Investement in new imported machines avoided.
h	Engineering	Automotive die cast component	Die casting	Automotive part	Histogram of dimension.	Dimensional rework reduced.

Reference	*Industry Group*	*Product*	*Process*	*Item*	*Approach*	*Problem/Benefit*
i	Engineering	Automotive part	Annealing	Automotive part	Histogram of hardness.	Lot one hundred percent inspection eliminated and rework reduced.
j	Plastic	Fibre	Filament formation	Fibre	Histogram of breaking load	Waste reduction through appropriate process control for each grade.
k	Engineering	Electroplated parts	Electro-plating	Plated parts	Histogram of bath parameters	Rework reduced, production enhanced, uniformity improved.
l	Fertiliser	Fertiliser	Packing	Fertiliser	Histogram of weight of filled bags.	Airconditioning avoided.
8.61 a	Engineering	Press shop	Drawing	Hemispherical bowl	Run chart of number non-conforming per sample.	Scrap of 10 percent eliminated.
b	Engineering	Job shop	Welding	Fabricated job	Run chart of non-conformities.	Assessment of quality and improvement for the first time.
c	Engineering	Job shop	Gas cutting	Fabricated job	Run chart of demerit score.	Demerit score improved from 35 to 5 per job.
d	Engineering electrical	Condenser plate	Drilling	Condenser plate hole	Run chart for hole diameter.	Non-conformity percent reduced from 8.0 to 0.3 percent.
e	Engineering electrical	Job shop	Machining	Turbine blades	Run chart (master) of non-conformities.	Non-conformities reduced from over 30 to under 0.5 percent and production enhanced four folds.
8.62 a	Chemical	Fertiliser	Grainding	Pulverised rock fineness	Run chart for residue on 100 mesh.	Variance reduced to one sixteenth.
b	Engineering electrical	Turbine blades	Machining	Turbine blades	Run chart on KR dimension.	Variance reduced to one twentyfifth.

Reference	*Industry Group*	*Product*	*Process*	*Item*	*Approach*	*Problem / Benefit*
c	Engineering	Automotive component	Machining	Automotive part diameter	Run chart on diameter.	(i) Potential to reduce variance to one tenth. (ii) Variance reduced to one fourth.
d	Engineering electrical	Rotor disc	Drilling	Rotor disc	Run chart on distance indicative of central location.	Standard deviation reduced to one twelveth (variance by 1/144).
9.1 to 9.8	Heavy electrical engineering	Turbine	Taping to curing	Turbo generator bar	Used all the seven tools.	Percent non-conformity reduced from about 5.0 to 0.5 percent.
10.2	Communication cable manufacturing	Telephone cable	Cable formation (impregnation)	Cable	Design of experiment.	Failure due to leakage reduced from about 95 to about 5 percent.
10.2	Insecticides	Disinfectant	Formulation and grinding	Disinfectant	Process control.	Rework reduced from thirty-five to one percent.
10.2	State transport	Service	Running	Mileage	Process control. (Speed and air pressure).	Fuel consumption reduced, tyre life increased.
10.2	Ceramic	Insulator	Pug milling and baking	Insulator	Process control.	Insulation property improved to acceptable level avoiding scrap.
10.2	Engineering	Component	Machining and assembly	Component	Process control.	Reduced rework on component and hiccup free assembly.

BIBLIOGRAPHY

Ackoff, R.L. and Rivett, Patrick, Manager's Guide to Operations Research. New York: John Wiley, 1963.

American Society for Quality Control; Wisconsin Quality Costs-What and How; Prepared by Quality Cost-Cost Effectiveness Technical Committee—Wisconsin: The American Society for Quality Control, 1971.

American Society for Quality Control, Wisconsin Guide for Reducing Quality Costs; Report of the Quality Costs Technical Committee—Wisconsin: ASQC, 1977, p. 46.

American Society for Quality Control, Quality Motivation Technical Committee, Quality Motivation Workbook—Wisconsin: ASQC Quality Pr., 1967, p. 126.

Asian Productivity Organization—Tokyo. (Industrial Engg. and Technology) Japan Quality Control Circles; Quality Control Circle Case Studies 1972 (4th Reprint, 1984) p. 208.

Bajaria, Hans J.; ed. Quality Assurance, Methods, Management and Motivation. Michigan: Society of Manufacturing Engineers, 1981, p. 248.

Bandyopadhyay, Jayanta K. QS-9000 Handbook; A Guide to Registration and Audit. Florida: St. Lucie Pr., 1996, p. 244.

Barker, Thomas B. Engineering Quality by Design; Interpreting the Taguchi Approach, New York: Marcel Dekker, 1990 i, p. 250.

Bieda, John, Practical Product Assurance Management. Wisconsin: ASQC Quality Pr., 1999, p. 249.

Biswas, Suddhendu, Statistics of Quality Control; Sampling Inspection and Reliability. New Delhi: New Age Intnl., 1997, p. 140.

Blazey, Mark L. Insights to excellence 1996; An Inside Look at the 1996 Baldrige Award Criteria. Wisconsin: ASQC Quality Pr., 1996.

Bossert, James L. Quality Function Development; A Practitioner's Approach—Milwaukee: ASQC Quality Pr., 1991, p. 127. (Quality and reliability V. 21)

Box, George E.P. and Others Statistics for Experimenters; An Introduction to Design, data Analysis and Model Building. New York, John Wiley, 1978. p. 653. (Wiley Series in Probability and Mathematical Statistics).

Boyd, Harper W. and Westfall, Ralph, Marketing Research; Text and Cases. 3rd ed. Illinois: Irwin, 1972, p. 813.

British Productivity Council, London, Quality and Reliability; Basic Principles Explained with 16 case Histories. London, p. 36.

Camp, Robert C. Benchmarking; The Search for Industry Best Practices that Lead to Superior Performance. Wisconsin: ASQC Quality Pr., 1989, p. 399.

Carter (Chuck), C.L.; Jr. Quality Assurance, Quality Control and Inspection Handbook/By C.L. Carter, Jr. and Associates. 3rd. ed., Texas: C.L. Carter, 1979.

Carter, C.L. (Chuck); Jr., The Control and Assurance of Quality Reliability and Safety. Texas: The Auth., 1978, p. 325.

Crosby, Philip B. Quality Without Tears; the Art of Hasele-Free Management. New York: McGraw-Hill, 1984, p. 205.

Croshy, Philip B. Quality is free; The Art of Making Quality Certain. New York: McGraw-Hill, p. 509.

Daniel, Cuthbert, Applications of Statistics to Industrial Experimental. New York: John Wiley, 1976, p. 294.

Das, M.N. and Giri, N.C. Design and Analysis of Experiments—New Delhi: Wiley Eastern, 1979, p. 295.

Deming, W. Edwards, Sample Design in Business Research—New York: John Wiley, 1960, p. 517.

Deming, W. Edwards, Out of the Crisis; Quality Productivity and Competitive Position—Madras: Productivity and Quality Publs., 1992 (Reprint 1993).

Deming, W. Edwards, Quality, Productivity and Competitive Position—Cambridge: MIT Pr., 1982, p. 373.

Drummond, Helga, The Quality Movement; What Total Quality Management is Really All About. London: Kogan Page, 1992, p. 160.

Feigenbaum, A.V. Quality Control; Principles, Practice and Administration—New York: McGraw-Hill, 1951, p. 443. (McGraw-Hill Industrial Organization and Management Series).

Ferber, Robert, Statistical Techniques in Market Research. New York: McGraw-Hill, 1949.

Ferber, Robert, Handbook of Marketing Research. New York: McGraw-Hill, 1974.

Flood, Robert Louis, Beyond TQM. Chichester: John Wiley, 1993.

Gilbert, John, How to Eat an Elephant; A Slice by Slice Guide to Total Quality Management. New Delhi: Affiliated East West Pr., 1992, p. 212.

Goldstone, Leonard, Statistics in the Management of Nursing Services. Kent: Pitman, 1980, p. 271.

Govil, A.K. Reliability Engineering—New Delhi: Tata McGraw-Hill, 1983, p. 330.

Greene, Richard Tabor, Global Quality; A Synthesis of the World's Best Management Methods. Milwaukee: ASQC Quality Pr., 1993, p. 886.

Greene, Richard Tabor, Global Quality; A Synthesis of the World's Best Management Methods. Milwaukee: ASQC Quality Pr., 1993, p. 886.

Hammer, Peter L. (Ivanescu) and Rudeanu, Sergiu Boolean Methods in Operations Research and Related Areas; Preface By Richard Bellman. Berlin: Springer-Verlag, 1968, p. 329. (Econometrics and Operations Research, V. 7).

Harrington, H.J. The Improvement Process; How America's Leading Companies Improve Quality—New York: McGraw-Hill, 1987, p. 239.

Harrington, H. James, Excellence; The IBM Way—Wisconsin: ASQC Quality Pr., 1988, p. 112.

Hemphill, Charles F. Management's Role in Loss Prevention. New York: American Management Assn., 1976, p. 186.

Hicks, charles R. Fundamental Concepts in the Design of Experiments—3rd Ed., New York: Rinehart and Winston, 1982i, p. 425.

Hutchins, David, Just in Time. England: Gower, 1988, xii, p. 215. Index..

Imai, Masaaki, Kaizen (Ky'Zen); The Key to Japan's Competitive Success—New York: McGraw-Hill, 1986, p. 259.

Indian Standards Institution, New Delhi, Sampling Inspection Tables. New Delhi: The Auth.

International Computers and Tabulators Limited, London, What is O.R.; An Introduction to Operational Research Techniques. London: The Auth., 1964, p. 23.

International Academy for Quality, The Best on Quality; Targets, Improvements, Systems. Ed. by John D. Hromi., Wisconsin: ASQC Quality Pr., 1996, p. 590.

Ishikawa, Kaoru, What is Total Quality Control? The Japanese Way—New Jersey: Prentice-Hall, 1985, p. 215. "Translated from the Japanese by David J. Lu".

Japanese Standards Association, Tokyo Guide to Quality Control and Company Standradization—Tokyo: Japanese Standards Association, 1984, p. 121.

Juran, J.M. Course on Management of Quality—4th. ed.—New York: United Nations, 1981.

Juran, J.M. Juran on Planning for Quality—New York: The Free Pr., 1988, p. 341.

Juran, J.M., Juran on Quality by Design; The New Steps for Planning Quality into Goods and Services—New York: The Free Pr., 1992, p. 538.

Juran, J.M., ed. A History of Managing for Quality, The Evolution, Trends, and Future; Directions of Managing for Quality; A Research Project Sponsored by Juran Foundation. Wisconsin: ASQC Quality Pr., 1995, p. 688.

Juran, J.M. Managerial Breakthrough; A New Concept of the Manager's Job, New York: McGraw-Hill, 1964, p. 396.

Juran, J.M. and Gryan, Frank M. Quality Planning and Analysis from Product Development Through Use. 3rd. ed. New York: McGraw-Hill, 1993, p. 634. (McGraw-Hill Series in Industrial Engg. and Management Science).

Juran, J.M. Juran on Leadership for Quality; and Executive Handbook. New York: The Free Pr., 1989, p. 376.

Kanji, Gopal K. and Asher, Mike, Hundred Methods for Total Quality Management. New Delhi: SAGE, 1996, p. 237.

Kanji, Gopal K. and Asher Mike, Total Quality Management Processes; A Systematic Approach. Oxfordshire: Carfax Publs., 1993, p. 132.

Kanti Swarup and Others, Operations Research/By Kanti Swarup, P.K. Gupta and Man Mohan—New Delhi: Sultan Chand, 1977, p. 728.

Karatsu, Hajime and Ikeda, Toyoki, Mastering the Tools of Q.C.; Learning Through Diagrams and Illustrations. Tokyo: PHP Instt., 1987, p. 243.

Karatsu, Hajime, TQC wisdom of Japan; Managing for Total Quality Control—Cambridge: Productivity Pr., 1980, p. 125.

Kehoe, Dennis F. The Fundamentals of Quality Management. London: Chapman and Hall, 1996, p. 289.

Labovitz, George and Others. Making Quality Work; A Leadership Guide for the Results Driven Manager/By George Labovitz, Yu sang Chang and Victor Rosansky. Essex: Oliver Wight Publns., 1992, p. 198.

Lester, Ronald H. and Others Quality Control for Profit/By Ronald H. Laster, Norbert L. Enrick and Harry E. Mottley. New York: Industrial Pr., 1977.

Logothetis, N. and Wynn, H.P. Quality Through Design; Experimental Design; Off-Line Quality Control and Taguchi's Contributions—Oxford: Clarendon Pr., 1989, p. 464.

Matsushita, Konosuke, Not for Bread Alone; A Business Ethos, A Management Ethic—Kyoto: PHP Instt., 1984 (Rep. 1987), p. 167.

Mears, Peter and Voehl, Frank, The Executive Guide to Implementing Quality Systems. London: Kogan Page, 1995, p. 218.

Milakovich, Michael E. Improving Service Quality; Achieving High Performance in the Public and Private Sectors. Florida: St. Lucie Pr., 1995, p. 258.

Miyashita, Kenichi and Russell, David, Keiretsu; Inside the Hidden Japanese Conglomerates. New York: McGraw-Hill, 1996, p. 225.

Mizuno, Shigeru; ed. Management for Quality Improvement; The Seven New QC Tools—Cambridge: Productivity Pr., 1988, p. 304.

Mizuno, Shigeru, Company-Wide Total Quality Control—Tokyo: Asian Productivity Organization, 1988, p. 313.

Moder, Joseph J. and Elmaghraby, Salah E., (Editors) Handbook of Operations Research. New York: Van Nostrand, 1978, p. 662.

Montgomery, Douglas C. Design and Analysis of Experiments—New York: John Wiley, 1976, p. 418.

Montgomery, Douglas C. and Peck, Elizabeth A. Introduction to Linear Regression Analysis. New York: John Wiley, 1982, p. 504.

Moss, Marvin A. Applying TQM to Product Design and Development. New York: Marcel Dekker, 1996, p. 412.

Moss, Marvin A. Designing for Minimal Maintenance Expense; The Practical Application of Reliability and Maintainability. New York: Marcel Dekker, 1985.

Munro-Faure, Lesley and Others, Achieving Quality Standards; A Step-by Step Guide to BS5750/ISO 9000/By Lesley Munro-Faure, Malcolm Munro-Faure and Edward Bones. London: Pitman, 1993, p. 210.

Murthy, M.N. Sampling Theory and Methods—Calcutta: Statistical Publs., 1967, p. 684 (Statistical Publishing Society; Books on Probability and Statistics)

Nelson, Charles, TQM and ISO 9000 for Architects and Designers. New York: McGraw-Hill, 1996, p. 317.

Oakland, John S. and Porter, Leslie J. Cases in Total Quality Management. Oxford: Heinemann, 1994, p. 271. Index, Appendix.

Oakland, John S. Total Quality Management; The Route to Improving Performance—2nd. ed.—Oxford: Heinemann, 1993, p. 463.

Odeh, Robert E. and Fox, Martin, Sample Size Choice; Charts for Experiments with Linear Models—2nd ed.—New York: Marcel Dekker, 1991, p. 194. (Statistics, Textbooks and Monographs; V. 122)

Park, Sung H. Robust Design and Analysis for Quality Engineering. Madras: Chapman and Hall, 1996, p. 329.

Patton, Joseph D., Jr. Maintainability and Maintenance Management—North Carolina: Instrument Society of America, 1980, p. 441.

Persico, John, Jr., ed. The TQM Transformation; A Model for Organizational Change—New York: quality Resources, 1992, p. 195.

Plowman, Brian, High Value, Low Cost; How to Create Profitable Customer Delight. London: Pitman, 1994, p. 267.

Poirier, Charles C. and Tokarz, Steven J. Avoiding the Pitfalls of Total Quality, Wisconsin: ASQC Quality Pr., 1996, p. 244.

Price, Frank, Right Every Time Using the Deming Approach/By Frank Price., Hants: Gower, 1990, p. 182.

Quality Management Library. CCTA, The Govt. Centre for Information Systems. London: HMSO, 1992. 1-Overview, p. 15, 2-Quality Management System Implementation/By C.J. Ashby, xv, p. 194, 3-Quality Management System Audit/By C.A. Moore, p. 94, 4-Quality Techniques/By C. Barratt, J. Slater and D. Morgan, ix, p. 75, 5-Quality Training/By C.A. Moore and C.J. Ashby, p. 61.

Rosander, A.C. Applications of Quality Control in the Service Industries—New York: Marcel Dekker, 1985, p. 387. Index, Bibl. (Quality and reliability)

Rosander, A.C. The Quest for Quality in Services. Wisconsin: ASQC Quality Pr., 1989, p. 568.

Ross, Joel E. Managing Productivity. Reston: Reston Publns., 1977, p. 191.

Rothery, Brain, ISO 9000 Aldershot. Hampshire: Gower, 1991, p. 182.

Russell, Robert S. and Taylor, Bernard W. Production and Operations Management; Focusing on Quality and Competitiveness. New Jersey: Prentice-Hall, 1995.i, p. 916.

Rust, Roland T. and Others, Return on Quality Measuring the Financial Impact of Your Company's Quest for Quality/By R.T. Rust, Anthony J. Zahorik and Timothy L. Keiningh. Chicago: Illinois Univ. Pr., 1994.

Sasaki, Naoto and Hutchins, David, ed. The Japanese Approach to Product Quality; its Applicability to the West. Collection of Papers Presented to the Past Four Conferences of QC Circle held in London since 1979. Oxford: Pergamon, 1984, p. 134.

Sayre, Don, Inside ISO 14000; The Competitive Advantage of Environmental Management. Florida: St. Lucie Pr., 1996, p. 232.

Seminar on Upper Management and Quality Juran Institute, N.Y. 1988, Upper Nanagement and Quality. 4th ed./By J.M. Juran. New York: Juran Instt., 1982.

Shecter, Edwin S. Managing for Worldclass Quality; A Primer for Executives and Managers. New York: ASQC Quality Pr., 1992.

Shewhart, W.A. Economic Control of Quality of Manufactured Product—New York: Van Nostrand, 1931, p. 501 (Bell Telephone Laboratories Series)

Shores, A. Richard Survival of the Fittest; Total Quality Control and Management Evolution. Wisconsin: ASQC Quality Pr., 1988, p. 302.

Sloan, M. Daniel, Using Designed Experiments to Shrink Health Care Costs. Wisconsin: ASQC, p. 215.

Snee, Ronald D. and Others Experiments in Industry; Design, Analysis and Interpretation of Results/ By Ronald D. Snee, Lynne B. Hare and J. Richard Trout Milwaukee. Wisconsin: ASQC Quality Pr., 1985.

Stebbing, Lionel, Quality Assurance; The Route to Efficiency and Competitiveness—3rd. ed.—New York: Ellis Horwood, 1993, p. 336.

Stein, Robert E. The Next Phase of Total Quality Management; TQM II and The Focus on Profitability. New York: Marcell Dekker, 1994.

Stephan, F.F. and McCarthy, P.J. Sampling Opinions; an Analysis of Survey Procedure—New York: John Wiley, 1958, p. 451 (A Wiley Publication of Related Interest to Statisticians)

Swanson Roger C. The Quality Improvement Handbook; Team Guide to Tools and Techniques. London: Kogan Page, 1995, p. 267.

Taguchi, Genichi and Others, Quality Engineering in Production Systems/By Genichi Taugchi, Elsayed A. Elsayed and Thomas Hsiang. New York: McGraw-Hill, 1989, p. 173.

Taguchi, Genichi, Introduction to Quality Engineering; Designing Quality into Products and Processes. Tokyo: Asian Productivity Organization, 1985, p. 191.

Taguchi, Genichi, On-line Quality Control During Production. Tokyo: Japanese Standards Asson., 1981.

Taguchi, Genichi and Wu, Yu-In, Introduction to Off-Line Quality Control. Tokyo: Central Japan Quality Control Asson., p. 115.

Tague, Nency R. The Quality Toolbox. Wisconsin: ASQC Quality Pr., 1996, p. 298.

Thomas, Brian, The Human Dimension of Quality. London: McGraw-Hill, 1995, p. 259.

United Nations. Industrial Development Organisation, Vienna Standardization Based on the Proceedings of the International Symposium on Development, Athens, Nov.-Dec. 1967. New York: United Nations, 1969, p. 66. (UNIDO Monographs on Industrialization of Developing Countries)

Wakhlu, Bharat, Total Quality; Excellence Through Organization; Wide Transformation. New Delhi: Wheeler Publns., 1994, p. 272.

Waller, Jenny and Others, The TQM Toolkit; A Guide to Practical Techniques for Total Quality Management/BY Jenny Waller, Derek Allen and Andrew Burns. London: Kogan Page, 1995.

Walton, Mary, Deming Management at Work. London: Mercury Bks., 1991, p. 252.

Wise, Stephen A. and Fair, Douglas C. Innovative Control Charting; Practical SPC Solutions for Today's Manufacturing Environment. Wisconsin: ASQC Quality Pr., 1998, p. 292.

Zaidi, A. SPC; Concepts, Methodologies and Tools. New Delhi: Prentice-Hall, 1995, p. 353.

Zairi, Mohamed, Total Quality Management for Engineers—New Delhi: Aditya Bks., 1992, p. 245.

OTHER FOREWORDS

I have gone through your book entitled "Problem solving Tools for Continuous Improvement" and your effort is appreciable give the subject. The explanation for the QC Tools has been covered exhaustively with adequate examples to clarify the concept and applications. *As far as my knowledge goes, there is hardly any other work/book, which covers this topic in so much details which reflects your rich experience in this field.*

I appreciate your efforts and wish you all the success and I hope that you will come out with many more such books and share your vast experience in the field to enrich the quality practitioners.

Chandra Shekhar

Sr. Director

It is indeed a matter of proud privilege and great pleasure to release the book on 'The Seven Magnificent, Simple, Quick and Cost Effective tools for Continuous Improvement' by Prof. A N Nankana.

What is Quality ? A question every professional in the field of quality management deliberated for many years and as always, has found customized answers. In other words, the answers that emerged out of quality tools and techniques would differ from the business environment under which 'Organisation' is bounded. The quality tools that are applicable to Far East (Japan) have been found not suitable for West (America) due to cultural differences in conducting business. In spite of such cultural differences there has been one area found where East meets West. This area is nothing but seven basic quality tools. Both East and West have found it very useful.

Although as a country, India has been Instrumental in revolutionizing quality at fundamental level, it found itself caught unawares, when it came to moving with the business world on 'quality management' area. It is sad that Indian Industry has been looking towards the West to validate the fundamental quality tools for a very long time. Thanks to liberalization, few Indian industries took plunge into the quality management area and found extra ordinary results from most fundamental and basic tools of quality. But such Indian case studies are in minority. No doubt these basic quality tools help entire Indian business, be it product centric or service specific. In the present day, to compete globally, it makes business sense to adopt, survive and progress before it is too late.

Prof. Nankana has been a pioneer in the field of quality and has spent a lifetime in raising the quality of Indian industry. The practice of simple statistical tools for continuous improvement has been one of the secrets of Japan becoming global quality leader. His book on Seven Magnificent Quality Tools is an extension of the same in Indian environment. It is an essential book for readers of every age, type and profession. The book is written with loads of Indian context and case studies from our own backyards that make it quite interesting. The book makes good reading and I am sure it will be of immense value to the readers.

M.K. Chari

Lt Gen

Director General Quality Assurance

Directorate General of Quality Assurance

Department of Defence Production

Ministry of Defence, Government of India

My association with Professor Nankana dates back more than forty years when he was advising us on the application of Statistical Quality Control methods in our Textile Mills. Thereafter, of course he helped us in our Engineering and Information Technology Divisions. My contact with him broadened while he was the Secretary of the Delhi Society for Quality Control of which I was the President.

Over the years, quality management has become an increasingly indispensable tool in all types of industry.

This book for which I have the honour to write a foreword is unique in its accessibility for the practitioner. It explains simple techniques through Cost and Effect Diagrams, Check Sheets, Parato Analyses, Stratification, Scatter Diagrams, Histograms and Run Charts in an exceedingly lucid manner. These basic approaches will be very helpful in dealing with almost every quality problem faced by industrial managers.

Furthermore, this book has presented an integrated way of looking at all aspects of the management of total quality which can be instrumental in continuous improvement. There are a number of case studies to illustrate and support the various quality control techniques. There is no doubt that the practice of these techniques will help towards achievement of various national and international awards.

Practising managers should look forward to possessing a copy of this book.

Dr. Vinay Bharat Ram
Chairman and Managing Director
The DCM Group, DCM Limited

I had the pleasure of having an extensive reading of the book "Seven Magnificent QC Tools" written by Prof. Nankana from the perspective of a student, a user in the industry as well as from the viewpoint of a trainer and assessor.

I feel that this is perhaps the first honest and diligent attempt made by an expert who has devoted his entire career for the promotion of a Quality Culture and make India Inc. brand shine across the globe.

Prof. Nankana has excellently put his life time experience of National and International Standardization, Industry Experiences as well his plethora of knowledge while being a teacher at Indian Statistical Institute in this book.

The book has been written in a very simple and lucid manner so that each one us especially in the field of Quality can keep this in our daily travel bag to our office / industry and use the same as a help book, text book and knowledge treatise for improvement practices.

I am sure that this magnificent piece shall be found extremely useful to all my colleagues in the Automotive, FMCG, Service Sector as well as in the Chemical and Food Processing Industries.

I wish Prof. Nankana many more years with us so that we can learn from his experiences and knowledge and truly put Indian Industry on to the Global Map.

B.B. Gupta
Regional Manager
U L India Pvt. Ltd.
(A Subsidiary of Underwriters Laboratories Inc. USA)

Kaizen or Continuous Improvement is the key in the ever-evolving work environment. However, the innovative ideas that spur the small as well as quantum improvements can be brought only on identification and understanding of the critical issues. The 'Quality' concept has also been evolving over the years with the spotlight on Customer Satisfaction and Delight attained through Continuous Improvement or Kaizen.

The continual improvement follows the renowned Deming's PDCA (Plan-Do-Check-Act) cycle. Checking or Diagnosis of the outcomes and outputs becomes the critical measure, which would guide us to take appropriate Preventive and Corrective actions. Various Statistical Tools are used in assessment and analyses of data, which may be related to the process parameters, customer satisfaction levels, attrition rates or other relevant indices. These statistical analyses help one to crystallize and formulate innovative means by directing the energies and efforts to the root causes. However, Statistics per se is viewed as a dry, dull and complex subject, tending to scare off many. Prof. Nankana brings to fore the 7 magnificent tools — viz. Cause and Effect Diagrams, Check sheets, Pareto analysis, Stratification, Scatter Diagram, Histogram and Run Charts — which are simple, cost effective and can be handy for practitioners in undertaking Kaizen initiatives. These tools are used in combination or independently to understand the root cause and magnitude of problems and to aid brainstorming for evolving Kaizen measures. These tools facilitate understanding of facts and ground realities, which is the foundation for analysis and Improvement proposal formulation. Such fact-finding is more relevant in countries like India, where we are often carried away by mistaken perceptions or prejudices rather than facts, resulting in confusion at implementation stage.

The strength of Japanese management lies in its capacity and emphasis on unearthing facts, which has helped them become icons of progress and innovation.

The literature available on Problem Solving Techniques often highlights examples of an alien environment. Prof. Nankana has in this book has explained clearly the 7 tools through vivid illustrations and examples, pertinent to Indian scenario from his vast experience; thereby demystifying the 7 magnificent tools, which can revolutionize Kaizen movement. I personally believe that these seven tools can work wonders in aiding the Problem Solving processes in the possessers of adept minds from all walks of life.

R.G. Monga
Deputy Director General
National Productivity Council of India

I am delighted to understand that as a part of your commitment to the mission of quality improvement for the Indian Industry, you are publishing your Book on 'THE SEVEN SIGNIFICANT TOOLS' for quick, simple, cost effective, quality control solutions.

In the fast growing Engineering Industry base in our country triggered by the exploding internal demand growth as also the growth of fast growing sourcing pull from the international markets, the Indian industry is in the midst of an environment of substantial opportunities for new business and faster growth.

However, to successfully convert these opportunities into sustainable business propositions, the industry needs to be more conscious now to produce high world-class quality of products and continuously improving their standards of performance.

Thus, your plans to publish your above book recognizes the above needs of the Indian Industrial Growth Environment. It effectively highlights the Seven Tools of Quality in such detail which are easy to understand and convincingly convenient to implement. The Book innovatively explains how useful a simple tool like 'stratification' can be to solve even some day-to-day issues and problems.

The principles are great, the language is simple and implementation convenient.

This Book can be readily understood and internalized not only at the management level but also at the operational level, individually and collectively.

I am happy, you have effectively put across in this book, the conclusions of your extensive research and wide consultancy experience.

Wish you all success.

I am sure this is yet another aspect of your commitment to serve the industry at this time when they are all engrossed in graduating themselves to the global platform.

With best wishes,
Yours sincerely,

V.P. Mahendru
Chairman–Managing Director
Past President–Indian Electrical and Electronics Manufacturers Association

To sustain in the globally competitive market one has to continually put in efforts for achieving excellence in quality. Quality is the sharpest weapon to survive, compete and prosper. It has been proved that statistical techniques are very helpful in improving and maintaining the quality.

This book presents magnificent seven tools of statistics in a very simple manner easy to understand and use. This book also gives case studies, which show how many industries have been benefited using these tools.

It is desirable that the benefits of statistical tools should be shared amongst wider readers. I am, therefore, happy Shri A.N. Nankana has taken a lead so as to enrich readers of his distinguished contributions in this field through this book. Prof. Nankana has been associated with Bureau of Indian Standards for more than three decades in preparation and promotion of National Standards on Statistical Quality Control and Quality Systems.

I wish Prof. Nankana all the best in this original venture and hope that readers will be immensely benefited from this book.

P K Gambhir
Scientist 'F' and Head (Training Institute)
National Institute of Training for Standardization
Bureau of Indian Standards
A-20 & 21, Sector 62
Noida

The focus of managing for quality has shifted from the quality of products and services (which is taken for granted) to what Juran calls the "big Q". Companies and organisations have to cope with ever increasing customer expectations along with lowering costs. Sustained growth with profitability can be achieved only by enhancing customer satisfaction by meeting customer requirements and through continual improvements.

Modern quality management practice is built around the well known eight principles. Of these, factual approach to decision making is of paramount importance. The need for and the ability to collect, analyse, interpret and present true, adequate and relevant data is essential for effective decision making at all levels from shop-floor to board room.

Effective use of data requires a good understanding of, and the ability to apply in practical situations, at least basic principles of statistics. No doubt a large number of books have been written on the application of statistical tools and techniques to quality control. Most of them are of western origin. Many practising engineers and technicians are overawed by the statistical content and treatment and consider these books to be of little use in their daily work management.

It is in this context that Prof. Nankana's "The 7 Magnificent Statistical Quality Control Problem Solving Tools for Continuous Improvement" provides a refreshing contrast.

This well written, jargon-free book reflects the distilled wisdom of Prof. Nankana's more than 4 decades teaching and 'hands-on' consulting practice covering a wide variety of industries. This rich experience has enabled the author to gain deep understanding of the mindset of practising engineers and technicians. This is clearly brought out by the precision of thought and lucidity of expression one finds throughout the book. The step-by-step development of concepts backed by illustrative examples from real life situations make the book eminently readable, by extremely useful for practising daily work management.

We in BHEL, right across all our factories located all over India including in doctrination of over suppliers and customers, have benefited by the applications of the tools and techniques. Prof. Nankana taught a very large number of engineers. We learnt how to convert a physical problem into a statistical problem, find a statistical solution and convert it into a physical solution. All the examples given in the book emphasise this fundamental approach.

Prof. Nankana calls SQC tools Simple, Quick and Cost Effective Tools. We whole heartedly agree, based on our experience.

I thank Prof. Nankana for the efforts he has put in to write this book among his continuing professional commitments.

New Delhi
September 2004

R. Subramanian
General Manager
BHEL, Corporate Quality